101 KEY IDEAS

Evolution

101 KEY IDEAS

Evolution

Morton Jenkins

TEACH YOURSELF BOOKS

For UK orders: please contact Bookpoint Ltd, 78 Milton Park, Abingdon, Oxon OX14 4TD. Telephone: (44) 01235 400414, Fax: (44) 01235 400454. Lines are open from 9.00–6.00, Monday to Saturday, with a 24 hour message answering service. Email address: orders@bookpoint.co.uk

For USA & Canada orders: please contact NTC/Contemporary Publishing, 4255 West Touhy Avenue, Lincolnwood, Illinois 60646–1975, USA. Telephone: (847) 679 5500, Fax: (847) 679 2494.

Long renowned as the authoritative source for self-guided learning – with more than 30 million copies sold worldwide – the *Teach Yourself* series includes over 200 titles in the fields of languages, crafts, hobbies, business and education.

British Library Cataloguing in Publication Data
A catalogue record for this title is available from The British Library.

Library of Congress Catalog Card Number: On file

First published in UK 2000 by Hodder Headline Plc, 338 Euston Road, London, NW1 3BH.

First published in US 2000 by NTC/Contemporary Publishing, 4255 West Touhy Avenue, Lincolnwood (Chicago), Illinois 60646–1975 USA.

The 'Teach Yourself' name and logo are registered trade marks of Hodder & Stoughton Ltd.

Copyright © 2000 Morton Jenkins

Cover design and illustration by Mike Stones

Typeset by Transet Limited, Coventry, England.
Printed in Great Britain for Hodder & Stoughton Educational, a division of Hodder Headline Plc, 338 Euston Road, London NW1 3BH by Cox & Wyman Ltd, Reading, Berkshire.

Impression number   10 9 8 7 6 5 4 3 2
Year   2005 2004 2003 2002 2001 2000

Contents

Introduction

Welcome to the **Teach Yourself 101 Key Ideas** series. We hope that you will find both this book and others in the series to be useful, interesting and informative. The purpose of the series is to provide an introduction to a wide range of subjects, in a way that is entertaining and easy to absorb.

Each book contains 101 short accounts of key ideas or terms which are regarded as central to that subject. The accounts are presented in alphabetical order for ease of reference. All of the books in the series are written in order to be meaningful whether or not you have previous knowledge of the subject. They will be useful to you whether you are a general reader, are on a pre-university course, or have just started at university.

We have designed the series to be a combination of a text book and a dictionary. We felt that many text books are too long for easy reference, while the entries in dictionaries are often too short to provide sufficient detail. The **Teach Yourself 101 Key Ideas** series gives the best of both worlds! Here are books that you do not have to read cover to cover, or in any set order. Dip into them when you need to know the meaning of a term, and you will find a short, but comprehensive account which will be of real help with those essays and assignments. The terms are described in a straightforward way with a careful selection of academic words thrown in for good measure!

So if you need a quick and inexpensive introduction to a subject, **Teach Yourself 101 Key Ideas** is for you. And incidentally, if you have any suggestions about this book or the series, do let us know. It would be great to hear from you.

Best wishes with your studies!

Paul Oliver
Series Editor

Acknowledgement

The author expresses his sincere thanks to Helen Green of Hodder & Stoughton for her help and encouragement throughout the production of this book. It is also a pleasure to thank Dr Sue Noake, Headteacher of Lewis Girls Comprehensive School, Ystrad Mynach, for her help and advice during the preparation of the text.

Adaptive radiation

Adaptive radiation is the gradual formation through evolution of a number of different varieties or species from a common ancestor, each adapted to a different ecological niche. The French scientist Georges Cuvier (1769–1832) laid the foundation to the conception of the idea because he realized that, over great periods of time, organisms had diverged into a number of different groups, some of which are still accepted as part of the classification system today. Such a view differed fundamentally from current doctrine, which was based on the *Scala Naturae* ('ladder of life') of Aristotle and envisaged all living things as occupying a single graded series from the simplest at the bottom to the most complex (i.e. mankind) at the top.

A classic example of adaptive radiation is illustrated by Darwin's finches. The beaks of the different species are adapted for diets associated with the particular ecological niches they occupy. They range from the formidable beaks of seed eaters and the curved beaks of flower feeders to the delicate, slender beaks of insectivores. The supposition is that this variety of well adapted species has been derived from a few more generalized ancestral colonists. Isolation, inherent variation, and competition for the available ecological niches must have been responsible for the patterns of adaptive radiation that were so clearly evident to Darwin. These were particularly striking among the reptiles and birds of the Galapagos Islands.

At a higher taxonomic level, fossil evidence suggests that the first primitive placental mammals appeared at the beginning of the Eocene period, 58 million years ago. Among these was the order Creodonta, which diverged into a number of families of carnivores, among them the marine mammals. These include the great whalebone whales with their highly specialized filter-feeding device, which enables them to strain off masses of small crustaceans (krill) from the sea water. Some of these whales attain a weight of 20 tonnes. The toothed whales, such as the sperm whale, feed largely on deep-sea squid, while the porpoises and dolphins are specialized for fish-eating.

see also...

Cuvier; Darwin's finches; Niche

Age of amphibians, The

It was in the Carboniferous period (360–286 million years ago) that amphibians flourished and spread. Many early types were much larger than those existing today. The harsh desert-like conditions of the Devonian period (410–360 million years ago) acted as an environmental selective pressure and forced restrictions on the basic shape of amphibians. They typically developed the four-limbed stance with the head above the ground and a barrel-shaped body holding the lungs. Climate changes and the consequent lush fertile swamps of the succeeding Carboniferous period provided a variety of habitats eminently suitable for the amphibian mode of life. As a result, the amphibious group spread by adaptive radiation and colonized the warm shallow waters of marshes as well as returning to a totally aquatic lifestyle – as we see in newts and some salamanders today.

It is likely that amphibians arose from several groups of lobed finned ancestors similar to the coelacanth because several types of Devonian lobe-fins have skeletal features that resemble those of later amphibians. Some had two pairs of lobed fins that were well placed for movement on land and others had nostrils similar to those of frogs. A fossilized skeleton of the earliest true amphibian was found in the 1930s in the upper Devonian deposits in Greenland and many major discoveries were made in the 1980s and 1990s. The fossil is called *Ichthyostega* and is one of the most perfect fossil links between major groups that has ever been found. However, unmistakable amphibian fossil footprints have been found in earlier Devonian deposits in Australia. *Ichthyostega* has a combination of fish and amphibian features. Its sense organs and long finned tail are more fish-like characteristics; its sprawling limbs and expanded ribs are more like those of amphibians. Although *Ichthyostega* is the oldest complete amphibian skeleton yet known, it is unlikely to have been the ancestor of today's amphibians as it was probably one of the many amphibian 'experiments' of the Devonian, most of which have ended as fossils with no evidence of surviving direct descendants.

see also...

Adaptive radiation; Coelacanth

Age of fishes, The

The end of the Silurian period (433–410 million years ago) saw a massive diversity in the fishes. This expansion coincided with their colonization of fresh water, which gave them problems opposite to those of living in the sea. Fresh water provided more environmental selective pressures than sea water because of its greater potential for variation in physical properties. It therefore provided greater opportunities for the development of species diversity. Temperature can change quickly, levels of nutrients can vary widely, and oxygen concentration can diminish rapidly. The marine environment tends to be more stable and less stressful in the context of these factors. The fishes have been defined as aquatic, cold-blooded, gill-breathing vertebrates which have retained fins rather than developing pentadactyl limbs. That is, they are backboned animals that live in water (salty and fresh water), cannot control their body temperature, and breathe oxygen dissolved in water by means of gills. Fishes first appear as fossils in marine sediments of Ordivician age, about 450 million years old. Since that time they have diversified to inhabit almost every aquatic environment. There are an estimated 23,000 species of fishes extant today; three out of every five vertebrate species are fish.

Acanthodians are the oldest known jawed vertebrates. *Climatius* is a well documented fossil of this group with thick, small scales all over its body. In contrast, members of the group Heterostraca, such as *Pteraspis*, and some ostracoderms were jawless fishes with a bony shield which enclosed the front part of their bodies. These are considered to be distant relatives of the lampreys and parasitic hagfish which are still alive today. The actual origins of fishes are poorly known because by the time the jawless fishes (Agnatha) and the jawed fishes (Gnathostomes) appear in the fossil record, they each already have the anatomical features that appear in modern forms today. However, it is considered that the Agnatha pre-date the gnathostomes, fossils of which are first found in mid-Silurian sediments.

Age of invertebrates, The

The Cambrian period (540–505 million years ago) is known as the Age of the invertebrates. Imagine that space shuttles, supersonic aircraft and all other modes of travel that human ingenuity could ever design appeared during a ten-year period in the fifteenth century. Today's historians would be queuing up to publish their explanations of how such a variety of creativity came about in such a short space of time. This is analogous to the attitude of palaeontologists to the explosion of evolutionary activity that took place in the Cambrian period. Late in 1993, scientists suggested that in just 10 million years living forms erupted in an orgy of innovation that far surpassed anything Earth had witnessed before or since. During the preceding 3,000 million years, the 'best' that evolution could come up with were the equivalent of dugout canoes in our transport analogy. There were algae, some flatworms and the mysterious Ediacara fauna. Then, without warning (at least little that left a fossil record) evolution burst forth like the proverbial bat out of hell, relative to the normal geological way of doing things. Early in the Cambrian period a large number of well defined invertebrate groups suddenly appeared. They were the first to have heads, middles, rear ends, segments and guts. Some had four legs, some a dozen. Blood, shells and antennae also first appeared. Indeed, almost all the body forms familiar to modern invertebrates, and several more that have long since disappeared, abruptly materialized in this one great release of evolutionary energy.

In the mid-1990s three scientists in California came up with an ingenious explanation for the arrival of so many diverse body plans in such a short time. They put forward the hypothesis that there appeared, at this time, cells that are held in reserve during larval development and which are later switched on to create the adult. These cells are switched on by homeotic genes. As soon as these evolved, there was the potential for the Cambrian explosion and the age of the invertebrates.

see also...

Ediacara fauna

Age of mammals, The

During the late Permian and early Triassic periods (about 245 million years ago), at about the same time as the ancestors of the dinosaurs were increasing their hold on the land, another group of reptile, the synapsids, was becoming dominant in many environments. They spanned a range of forms, from the more primitive reptiles to creatures very similar to mammals – the therapsids.

An early innovation among the synapsids was the evolution of a rather clumsy temperature control mechanism. Some species developed a large web of skin along their backs: a 'sail', supported by long extensions of the vertebrae. The large surface area of the sail, richly supplied with blood vessels, would have enabled these 'sailbacks' to absorb the sun's energy when they were cold, and lose heat when they were too hot. A constant body temperature permitted the synapsids greater activity, regardless of the external temperature, and was an important step in the development of temperature-dependent enzyme and hormone systems geared to a regular high metabolic rate; a feature common to the mammals that may have evolved from them. Some synapsids, which could have been ancestors of mammals, evolved more efficient methods of locomotion, feeding and breathing.

These mammal-like reptiles show an almost perfect series of transitions from the reptilian to the mammalian condition. Moreover, it is likely that many animals now considered on purely skeletal evidence to be reptiles would be called mammals if they were alive today. The arbitrary point at which the two classes are separated concerns the structure of the jaw. The skeletal definition of a mammal is that only one bone is present on each side of the lower jaw. If this strict definition is applied to fossils, creatures already present in the middle Triassic period must be considered to be mammals. By the middle Jurassic period (about 190 million years ago), there were various groups of small mouse-like mammals, something like modern shrews. These early mammals diversified quickly even during the reign of the dinosaurs. Many became extinct, but enough survived to inherit the post-dinosaur world.

see also...

Dinosaurs

Age of reptiles, The

The evolution of some ancestral amphibians into reptiles took place in the Carboniferous period (360–286 million years ago). By the time the luxurious coal swamps – the Eden of the amphibians – had given way to the bleak deserts of the Permian (286–245 million years ago), and the Triassic (245–202 million years ago), the reptiles had become total masters of the land and were to remain so for the next 200 million years.

The oldest known reptile was thought to be *Hylonomus*, from the mid-Carboniferous rocks of Nova Scotia, Canada (about 310 million years old). This small animal, 20 cm (8") long, probably looked like a lizard. The skull bones and the general proportions of the body show that it was an active land animal – a member of a new group that had progressed beyond the land-living amphibians in its adaptations. *Hylonomus* was discovered by William Dawson in 1852. His record of finding the remains of the world's most ancient reptile lasted for nearly 140 years. Then, in 1989 an important discovery was made by the Scottish fossil collector, Stan Wood. He extracted a fossil reptile from the early Carboniferous rocks of Scotland, and pushed back the date of the origin of reptiles closer to the origin of the amphibians (350 million years ago).

After a modest beginning, the reptiles came to dominate the Permian period. In particular, the mammal-like reptiles were the first major group to exploit the possibilities of a fully terrestrial existence. They included the famous sail-back reptiles of the early Permian, and the diverse and successful carnivorous and herbivorous forms of the late Permian. Some of these were the first animals to achieve large size. Other innovations include the conquest of the sea by the mesosaurs, and the first flying vertebrates, the pterosaurs. All the major reptilian lineages came on the scene during the late Carboniferous period, even though the amphibians still seemed to be the dominant vertebrates in most parts of the world.

see also...

Cleidoic egg; Dinosaurs; Pterosaurs

Allopatric speciation

The ways in which species arise are dependent on geography. Most species arise through allopatric speciation (*allo* = 'other'; *patric* = 'land'), which means the formation of new species after the geographic separation of once continuous populations. The process involves a population being somehow divided and each subgroup taking a different evolutionary route until they have diverged so much that interbreeding is no longer possible, even if they should rejoin. A species has been defined by the renowned zoologist Ernst Mayer, as follows:

> A species is a group of actually or potentially interbreeding populations that is reproductively isolated from other such groups.

Small emigrant populations of any sort may show all the rapid changes which are associated with limited size, such as genetic drift and the founder effect. They are also protected against gene flow from the parent population.

Populations can be divided in two basic ways:

★ One is a small group of individuals (or even a seed of a plant or a pregnant animal) being separated from the parent population and the descendants therefore becoming established in a new place. An example is Darwin's finches – they became genetically isolated not only from the parent population but also from each other as the various islands gave rise to their own distinct forms.

★ Populations have also been divided by geographical events, perhaps as some great barrier appeared to separate a group into two smaller groups that then went their own evolutionary ways. An example is seen in the Abert and Kaibab squirrels. These are two distinct species that live on opposite sides of the Grand Canyon. It is believed that they were once one population that was divided as the great chasm developed.

see also...

Darwin's finches; Founder effect; Genetic drift; Geographic isolation; Reproductive isolation; Sympatric speciation

Allopolyploidy

A type of polyploid mutation involving the combination of chromosomes from two or more different species. Allopolyploids usually arise from doubling of chromosomes of a hybrid between two species, the doubling often making the hybrid fertile (this is called amphiploidy). The properties of the hybrid, e.g. greater vigour and adaptability, are retained in the allopolyploid in subsequent generations and such organisms are often highly successful.

The most commonly encountered type is the allotetraploid, which has two complete sets of genes from each of the two original parent species. The Russian cytologist Karpechenko (1928) synthesized a new genus from crosses between vegetables belonging to different genera, the radish (*Raphinus*) and the cabbage (*Brassica*). These plants are fairly closely related and belong to the same family (*Cruciferae*), which includes mustard. Each has a diploid chromosome number of 18, but the chromosomes in the radish have many genes that do not occur in the cabbage, and vice versa. Karpechenko's hybrid had in each of its cells 18 chromosomes, nine from the radish and nine from the cabbage. Members of the very dissimilar sets of genes failed to pair during meiosis and the hybrid was largely sterile. A few 18-chromosome gametes were formed, however, and a few allotetraploids were therefore produced in the F_2 generation. These were completely fertile, because two sets each of radish and cabbage chromosomes were present and pairing between homologous chromosomes took place at meiosis. The allotetraploid, or amphidiploid, was named *Raphanobrassica*. Unfortunately, the resulting plant had the root of a cabbage and leaves of radish and so is of no economic value. The method does, however, illustrate the principle of producing fertile interspecific or intergeneric hybrids.

At least half of all naturally occurring polyploids are allopolyploids. Cultivated wheat provides a good example. It has 42 chromosomes, representing a complete diploid set of 14 chromosomes from each of three ancestral types.

see also...

Mutation

Antibiotic resistance

Antibiotic resistance in bacterial pathogens poses serious health problems, particularly in parts of the world where antibiotics can be freely purchased from pharmacies. The typhoid epidemic of 1972 in Mexico was an infamous reminder of this: under tragic circumstances, there was found to be widespread resistance by the typhoid bacillus to chloramphenicol. As an evolutionary phenomenon, antibiotic resistance arises basically by the selection of preadapted resistant mutants. Since the early 1950s it has been known that it is possible to select for resistance and to produce pure cultures of resistant strains which have never been in contact with the antibiotic.

The genes and cellular mechanisms involved in resistance vary between the different antibiotics and between species, and this is reflected in the rate at which resistance can be selected for. However, even within species, the mechanism of resistance to a given compound may be variable. Penicillin resistance has been recognized since the 1940s and arises because of the action of a bacterial enzyme which can break down penicillin (most naturally occurring penicillin-resistant strains of bacteria owe their resistance to this enzyme). Resistance to antibiotics has already destroyed or limited the usefulness of several formerly valuable drugs and selection of resistant bacteria in the presence of the drug can be rapid. Recent evidence of this is seen in the build-up of antibiotic resistance in several strains of tuberculosis-causing bacteria.

Superimposed on this problem is another, which seems to be unique to bacteria. Some strains can transmit the genes controlling resistance to others during a process called conjugation among sexually active individuals. This discovery was first made in 1959 by Japanese scientists, who demonstrated that the genes are transferred on strands of DNA like a rope being passed between two 'mating' strains of bacteria. The discovery of transferred resistance caused reappraisal of the use of antibiotics in animal husbandry, where they were employed not only to cure infectious diseases but also in attempts to prevent infections, and in some cases as food additives to promote the growth of the animals.

Archaeopteryx

There is no doubt that birds evolved from now extinct types of reptiles. Until the 1990s, palaeontologists were certain that the earliest bird was *Archaeopteryx lithographica*. The name means 'ancient wing from the lithographic stone' and is derived from *Archaeo* = 'ancient', *pterxy* = 'wing', and *lithographica* = a fine-grained limestone used for a particular type of printing called lithography. When alive, it would have had a mass of about 270 g and a wing area of about 480 cm^2 – about the size of a magpie.

Archaeopteryx had a long lizard-like tail and beak-like jaws lined with teeth, but its collar bones were united to form a bird-like 'wishbone' and it had feathers almost identical to those of modern birds. It dates from the Upper Triassic, some 140 million years ago. A fossil feather was discovered in 1860 in the fine-grained limestone from Solnhofen, Bavaria, and it was given a scientific name a year later by Hermon von Meyer. In the same year, in the same area, an almost complete fossilized skeleton was found. This fossil eventually found its way to the Natural History Department of the British Museum, having been purchased for the then princely sum of £700 by the curator, Richard Owen.

A second, more complete, specimen was found in 1877 and purchased by the Humbolt Museum, where it still resides (it is called the 'Berlin specimen'). It seems that during the Jurassic (202–144 million years ago), muddy lagoons were common in the Solnhofen region of Bavaria and it was there that the ancient birds became trapped and eventually fossilized. The result of 140 years' searching after the first find has been the discovery of seven fossils of *Archaeopteryx*. Then, in 1994, farmers uncovered hundreds of fossils of early birds in the Liaoning province of China. One of these was named *Confuciusornis sanctus* 'the holy Confucius bird', which could be older than *Archaeopteryx*. In 1996, another fossil bird, *Protoarchaeopteryx*, was found in the same area, which threw doubt on *Archaeopteryx* being the ancestor of modern birds. These newly found fossils could predate *Archaeopteryx* by 30 million years.

see also...

Owen

Artificial selection in animals

For thousands of years selected pedigree animals have been prized. Such animals have a recorded ancestry for many generations. Selective breeding aims to produce animals progressively better suited to human needs: the dairyman wants selected pedigree cattle from which to breed milking cows; the hunter wants a pedigree dog. Until relatively recent times, this form of breeding by artificial selection was somewhat hit or miss. Before the work of Gregor Mendel in the 1860s, no one had a clear idea of even the most basic principles of heredity, so the development of a new strain of plant or a new breed of animals was by trial and error. Ever since humans began to domesticate animals, hundreds of different varieties of animal breeds have gradually been developed: all modern chickens have been selectively bred from the jungle fowl of the Far East; the ancestor of the pig is the wild boar; that of the cow, the wild ox, and that of the over-weight Christmas turkey, the wild and slender flying turkey hunted by the original pilgrims of New England.

An example showing the value of selective breeding concerns the cattle of India. For centuries the typical Indian cow has been an unkempt and semi-wild creature. Breeding was haphazard and milk production pitifully small. You can still find such animals in existence today. The cow is considered sacred by Hindus and roams the streets of Indian villages unmolested and unattended, picking up food as it goes along. On government farms an attempt was made to improve these animals. The work began by giving them better food, but this did not significantly improve their milk yield. The problem was that the cows, being nervous creatures, are easily frightened and do not relax enough to yield very much milk when milked by hand. Eventually, after several years of selective breeding, a breed of less excitable cattle was produced, which allowed people to hand milk them. After 20 years, milk production was increased threefold. Today, a knowledge of selective breeding is combined with artificial insemination to increase the production of desirable breeds.

Artificial selection in plants

One of the first pioneers of selective breeding in plants was the American, Luther Burbank. He was responsible for developing a huge range of improved varieties of plants. Perhaps his most famous contribution to plant breeding was his success with potato crops. One day in 1871, while examining a field of potatoes in Massachusetts, Burbank noticed fruit growing on one of the plants. Although potato plants normally have flowers, they seldom produce fruit. New plants tend to be grown from potato stem tubers rather than from seeds. He saved the seeds and planted them. He checked the potatoes growing on the resulting plants and saw that they differed from plant to plant. Some were large, others were small; some produced many potatoes, others produced few. One plant had more potatoes that were also larger and smoother than the others. He selected this plant for future breeding by asexual means. It was named the Burbank variety of potato, and soon became popular throughout the USA.

Burbank's success was an example of mass selection, where one plant is chosen for breeding from a larger number of individuals. Mass selection is probably the oldest type of artificial selection. The first people to cultivate plants always saved the seeds for the next planting from plants that produced the best yield. The offspring are then most likely to have the desired traits and, over countless generations, modern crops have been evolved from their wild ancestors. Cereals, for instance, have been developed from wild grasses. Cauliflowers, broccoli, cabbage, and Brussels sprouts all belong to a family which originated as wild plants near sea shores.

Mass selection can also produce strains of disease-resistant plants. For example, suppose a fungus spreads throughout a wheat-producing area. Almost all the wheat is killed, except for two or three plants. Seeds from these healthy plants are grown in the following year. Again the fungus attacks the crop, but this time, more plants survive. Over several years this cycle is repeated and each year more plants survive to produce a fungus-resistant strain of wheat.

Autopolyploidy

A type of polyploid mutation involving the multiplication of chromosome sets from only one species (*auto* = self; *poly* = many; *ploid* = chromosome number). Autopolyploids may arise from the fusion of diploid gametes of the same species that have resulted from chromosomes failing to separate at meiosis (the type of cell division which produces sex cells). Alternatively, like allopolyploids, they may arise by the failure of chromatids to separate during the division of a fertilized egg. The hybrid formed as a result of autopolypoloidy may be fertile or sterile, depending on the number of chromosome sets. Hybrids with an even number of homologous chromosome sets (4, 6, 8, ... 28) will be fertile because chromosome pairing is possible at meiosis.

When a gamete, unreduced at meiosis, remains diploid (2x) it may fuse with a normal haploid one (x) giving a triploid offspring (3x). The union of two unreduced (2x) gametes gives an autotetraploid. Triploids can also arise from crossings between diploids and tetraploids. Another source is spindle failure at mitosis, which gives direct doubling of the somatic chromosome number and leads to the production of a polyploid cell. Derivatives of the doubled up cell could be 4x, 6x, or 8x, and may then form a polyploid part like a branch in an otherwise 2x, 3x, or 4x plant. Autopolyploids have chromosome sets that are all homologous with one another.

Chrysanthemum, the celandine (*Ranunculus ficaria*), the hyacinth (*Hyacinthus orientalis*), and some varieties of Cox's orange pippin (*Malus pumila*) represent this type of polyploidy. Artificially induced autopolypoloidy, with the use of colchicine, has been used to produce new and more vigorous varieties of commercial crop plants such as tomatoes and sugar beet.

see also...

Allopolyploidy; Mutation

Biogenesis

The hypothesis that living things come only from other living things is called biogenesis. It is now taken for granted by all biologists but it took 300 years for it to be accepted in favour of abiogenesis. In the eighteenth century, those early scientists who believed in abiogenesis and those who believed in biogenesis both used the same experimental material to prove their respective points. They used hay infusion, which was readily available, easy to prepare and, with the help of increasingly more sophisticated microscopes, easy to observe. Chopped hay was boiled for 10 minutes in water and left exposed to air. For the first few days, the liquid was clear and was observed under a microscope to be free from microbes. After a few more days, however, the liquid became cloudy and teemed with life. Both abiogenesists and biogenesists agreed that boiling the hay at the start would kill any living thing that might be present. Therefore, they concluded that the living creatures must have developed after the hay had cooled.

The abiogenesists explained that the living things were generated from the hay and water. To these, the biogenesists' explanation, that the air contained spores which contaminated the liquid after it had cooled, seemed incredible. The biogenesists suggested that the spores changed into active creatures. Pure air must be a complex mixture and every time people took a breath, they inhaled masses of strange creatures! The burden of proof was with the biogenesists.

The biogenesists deduced that, if they prevented air from coming in contact with the boiled, cooled hay, microbes would not appear. As long ago as 1711, Louis Joblot (1645–1723) tested the hypothesis but it was left to one of the greatest biologists of all time, Louis Pasteur (1822–1895), to finally prove the hypothesis of biogenesis. He demonstrated that:

★ all observations and experiments thought to be examples of spontaneous generation were false
★ in no experiment could it be shown that spontaneous generation could occur.

see also...

Spontaneous generation

Biogeographical evidence

By studying the present distribution of members of a group as well as the distribution of their fossil ancestors, we can gather a great deal of evidence for the evolution and subsequent spread of animals and plants throughout the world. There are many examples but a well documented one, the camel family, will serve to illustrate the principle. The llama and its relatives, the vicuña and the guanaco, are found in South America. The habitats of true camels are in North Africa, Arabia and Asia. They have gradually spread over these areas from their point of origin.

While only two genera, *Lama* and *Camelus*, survive today, in past ages there were many more. At least 25 fossil genera have been identified. Originating in the Upper Eocene (about 50 million years ago) all but one of those genera lived in North America. The only fossil camels found outside North America are of Pleistocene date (2 million years ago) and belong to our two living genera or to *Paleolama*, an extinct South American llama form. Despite the modern distribution of the camel family, clearly North America was the central point of its origin before they crossed to Asia to evolve into the Bactrian and Arabian camels, and South America to evolve into the llama, vicuña and guanaco.

Marsupials illustrate the same principle. The evidence indicates that they originated on the South American and Australian continents, which were once connected. Australia separated from all other land masses before the modern placental mammals gained a foothold.

Theatres of evolution need not be the size of whole continents, but for populations of animals or plants to spread from one great land mass to another millions of years are often required and communicating pathways must be present. These have changed during continental drift. The emigrating group must somehow cope with water barriers, high mountain ridges and zones of hostile climates.

see also...

Adaptive radiation; Comparative anatomy; Continental drift; Fossil evidence; Natural selection; Recapitulation evidence

Bird-hipped dinosaurs

The 'bird-hipped' dinosaurs are called orthischians. They have hip girdles in which the two lower bones on each side lie parallel and point backwards. They were all plant eaters and, as they spread during the Jurassic and Cretaceous periods (202–65 million years ago), were preyed upon by the carnivorous 'lizard-hipped' dinosaurs (saurischians) that developed at the same time.

The earliest ornithischians were quite small and two footed. Their bipedal stance developed together with the long tail and long hind legs of their archosaur (ruling reptiles) ancestor, in the same way as the two-footed stance of their lizard-hipped cousins. The ability to run on two legs would have given them greater speed and would have enabled them to escape from the swift meat eaters.

The bird-like arrangement of the hip bones in ornithischians allowed for a larger volume of internal organs than that of the saurischians. This difference is because a plant eater's digestive system occupies far greater volume than that of a meat eater. The early ornithischians were both bipedal and herbivorous, with the weight of the body still balanced at the hips. In contrast, plant-eating saurischians had bodies that extended forward, beyond the hips, to accommodate the increased volume of the intestine; hence the four-footed stance of the gigantic sauropods.

One group of primarily bipedal ornithischians developed specialized teeth to grind up vegetation. These were the ornithopods, or bird-footed types. The most successful of these were the duck-billed dinosaurs of the Cretaceous. Some reached a length of 12 m. Their skulls were expanded into a toothless beak at the front and contained thousands of closely packed teeth forming a grinding surface at the back.

Ornithischians also developed four-footed forms. These were often highly armoured and had shapes that resemble modern-day rhinoceroses.

see also...
Dinosaurs; Lizard-hipped dinosaurs

Bottle-neck effect

opulations may sometimes be reduced to low numbers through periods of seasonal climatic change, heavy predation, disease or catastrophic incidents involving volcanic eruptions or other natural disasters. As a result, only a small number of individuals remain in the gene pool to contribute their genes to the next generation. The small sample that survives will often not be representative of the original, larger, gene pool and the resulting allele frequencies may be severely altered. In addition to this 'bottle-neck' effect, the small surviving population is also often subject to inbreeding and genetic drift.

In a nutshell, the life history of a population subjected to a bottle-neck is:

★ a large population exists with very much genetic diversity
★ the population crashes to a very low number and loses nearly all of its genetic diversity and becomes almost extinct
★ the population grows to a large size again but has lost much of its genetic diversity.

The principle has been illustrated by studies of the world population of cheetahs. In recent years, the world population of wild cheetahs has declined to fewer than 20,000. Modern techniques of genetic profiling have shown that the total cheetah population has very little genetic diversity. Cheetahs appear to have narrowly escaped extinction at the end of the last ice age, 10–20,000 years ago (perhaps a single pregnant female survived in a cave and produced a litter). All modern cheetahs seem to have arisen from this one surviving litter – accounting for the lack of diversity. It has been estimated that, during the last ice age, 75 per cent of all large mammals died out (including mammoths, cave bears and sabre-tooth tigers). The lack of genetic variation in cheetahs has led to sperm abnormalities, decreased fecundity, high cub mortality and sensitivity to disease.

Since the genetic bottle-neck, there has been insufficient time for random mutations to produce new genetic variations.

see also...

Genetic drift; Genetic profiling

Broom, Robert

Dr Robert Broom (1866–1951) was an exceptionally skilful scientist who made important contributions to palaeontology, particularly relating to reptiles and humans. Broom was a Scot who in 1925 was working as a country doctor in a series of small South African outposts. He also had a brilliant and well-respected career as a palaeontologist, having been awarded a medal by the Royal Society in London for his research.

In 1934, at the age of 68, Broom retired from his medical career and took a post at the Transvaal Museum in Pretoria. He was encouraged to search for evidence of *Australopithecus africanus* ('African ape of the south'). In the 1920s, fossils of this creature had been found in limestone quarries and Broom decided that he should look there. One Sunday in 1936, he drove out to examine the limestone caves and quarry at Sterkfontein, near Johannesburg. He learned that the quarry manager, a Mr Barlow, had once worked in an area where human ape-like fossils had been found, and that he was saving fossils to sell to tourists. The owner of the quarry wrote in a little guide book to the places of interest near Johannesburg 'Come to Sterkfontein and find the Missing Link'. A strange prophecy indeed, because Broom did just that! On 17 August 1936, Barlow handed him a fine brain cast which turned out to be that of *Australopithecus*. During that day and the next Broom found the impression of the top of the skull as well as its base, with parts of the forehead and side walls.

His undoubted pleasure at the time of this discovery is summed up by this quotation:

> To have started to look for an adult skull of *Australopithecus*, and to have found an adult of at least an allied form in about three months was a record of which we felt there was no reason to be ashamed. And to have gone to Sterkfontein and found what we wanted within nine days was even better.

The amazing Dr Broom died in 1951, aged 85, still an enthusiastic fossil hunter.

Burgess Shale,

The Burgess Shale is a particular rock formation which has provided a harvest of unique fossils, few of which resemble any living forms. The Shale lies in the Yoho National Park on the slopes of Mount Stephen in British Columbia, Canada. Burgess was a nineteenth-century governor general of Canada and gave the name Burgess Pass to an access route to a local quarry from the town of Field. It has been well known by palaeontologists since 1909 when it was first made famous by Charles Doolittle Walcott, who found vast numbers of remains of animals that had been covered by sediments so quickly that even their soft parts were preserved for posterity as fossils.

Legend has it that while Walcott was exploring the area, his mule lost a shoe, forcing him to stop at a place where he noticed something glistening on the surface of a black slab of shale. The rock was of the Cambrian period (540–505 million years old) and on further examination it revealed a collection of fossils of large organisms hitherto unknown to science. Walcott traced the origins of the slabs of shale to the hillside, from where hundreds of specimens were later collected and kept in the United States National Museum in Washington, DC. The fossils appeared to be so bizarre that they were worthy of international interest and so many samples were sent to experts around the world for identification. Most spectacular were the fossil arthropods – the jointed-legged invertebrates – of which the trilobites had once been regarded as almost the only Cambrian types. This idea proved to be totally wrong because the shales produced an amazing variety, indicating that the oceans of that time were teeming with masses of tiny swimming crustacean-like animals.

In his best-selling book *Wonderful Life*, Stephen Jay Gould focuses on the evolutionary significance of this collection of strange fossils in meticulous detail. Film buffs will recall Frank Capra's 1946 movie *It's a Wonderful Life*, which gave Gould his title and presented the idea of replaying life stories.

Life on Earth has come very close to death no fewer than five times in the past when most of the living organisms on Earth suddenly became extinct. Indeed, we may be in the middle of the sixth mass extinction at present. However, the next mass extinction may be a result of self-inflicted destruction rather than being triggered by an external force.

The global ecosystem has suffered many times in its 3500 million year history and in the worst of these crises life came very close to total extinction. About 250 million years ago, life was on a knife edge in terms of survival. In a relatively short geological time span, at least 96 per cent of species died out. Of all the species that have ever lived on Earth, 99.9 per cent are now extinct. Often it seems that their demise was hastened by major changes to the planet. The causes and effects of these changes included numerous asteroidal and cometary impacts or near misses, producing huge dust clouds and climatic extremes changing Earth from a scorching planet to an ice box and back again. There have been major changes in sea level and prolonged periods of volcanic activity caused by moving tectonic plates.

The five major catastrophes were:

★ Late Ordovician (440 million years ago) – extinction of about 85 per cent of species.
★ Late Devonian (365 million years ago) – many marine species became extinct in two waves, one million years apart.
★ Late Permian (251 million years ago) – extinction of about 96 per cent of species. This was the largest mass extinction and nearly annihilated the mammal-like reptiles that had ruled for 80 million years.
★ Late Triassic (205 million years ago) – extinction of approximately 76 per cent of (mainly marine) species.
★ Late Cretaceous (65 million years ago) – extinction of between 75 and 80 per cent of species. This is the most infamous mass extinction of all because it heralded the end of the reign of the dinosaurs, which had dominated the land for 140 million years.

see also...

Continental drift; Dinosaurs

Cladistics

ladistics is a method of describing evolutionary pathways through the science of classification (taxonomy). One school of taxonomy subscribes to a scheme originally attributed to the German entomologist Willi Hennig in the 1950s and supported by Niles Eldredge and Joel Cracraft. It is based on the idea of genealogy, which attempts to record descent from ancestors via pedigrees. Hennig coined the term cladistics (derived from the Greek word, *klados* = 'branch' or 'shoot'). The basic group or taxon is the clade and organisms grouped within it are assumed to have been derived from the same ancestor. An important assumption in the construction of pedigrees (cladograms) is that all species present at a particular time are not themselves ancestors of other existing species. It follows, therefore, that they must be related, if somewhat distantly by descent from other ancestral forms.

Cladograms are constructed on the basis of a number of shared characteristics which are considered homologous and therefore indicate an evolutionary relationship. Moreover, the similarities must exist at the assumed branching point of each clade. Cladograms, then, are maps which show the relationships between species as a series of regularly splitting branches like a family pedigree showing your ancestry. Indeed, the word 'pedigree' comes from the French for 'crane's foot', *pied de grue*, due to its resemblance to a foot with toes branching from it.

Every branching point is where a new character was acquired, which is handed down to any species branching from it. Names of groups of animals or plants are given to clades that share a set of common characteristics. The groups can be very large – a good proportion of the animal kingdom – or very small, such as a genus comprising a handful of species, but the principle is the same. When constructing a cladogram, one of the greatest problems is to determine whether the characters used for comparison are really homologous between different species (sometimes known as derived characters). If similarities within a cladist group are shared with an organism outside it, these are probably not derived and are referred to as primitive characters.

Cleidoic egg

The evolutionary leap between the amphibians and the reptiles was possible only because of the development of a special type of egg that could be laid on land (the cleidoic or amniotic egg).

This feature is the main criterion distinguishing amphibians from reptiles. It permitted reproduction to occur without dependence on water and therefore allowed vertebrates to radiate away from aquatic habitats. Firstly a tough shell was laid down around the egg; this prevented desiccation. Secondly, large supplies of yolk were provided, which support growth and allow the young reptile to emerge from its egg as a fairly large and active animal. Thirdly, a system of membranes inside the egg provided both a miniature 'pond' in which the embryo could develop and a storage sac for excretory products.

Superimposed on these developments was the evolution of a form of excretory physiology different from that of totally aquatic animals. Those invertebrates and fish which live permanently in water can produce toxic ammonia as an excretory product. It is made from excess amino acids

from protein intake in a relatively primitive biochemical pathway. The ammonia is diluted by the large volume of water in which these animals live. With the development of eggs that can be laid on land, it would be impossible for the embryos within the egg to survive if they produced toxic ammonia – an accumulation during their development within an egg shell would be fatal. In order to overcome this, a more complex biochemical pathway evolved to produce harmless, insoluble uric acid instead of ammonia, which can be stored until the egg hatches.

Other consequences of the development of the cleidoic egg:

★ Elimination of a gill-breathing larval stage. Reptiles emerge from their eggs as mini-adults and no metamorphosis (as in amphibians) is needed.

★ Development of a method of internal fertilization. Sex cells can no longer be freely released into the water because fertilization has to be effected internally before the relatively impermeable shell is deposited around the egg in the oviduct.

Coelacanth

In 1938, an amazing discovery was made when a coelacanth was netted off the Comoros Islands. This catch of the century was equivalent to finding a living dinosaur. The fossilized remains of coelacanths are known in sedimentary rocks between 400 million and 65 millions years old but their fossil record mysteriously ends at the time of the demise of the dinosaurs. The fish are large, bluish grey with white blotches and have unusual stubby, lobed fins. The first specimen was landed at East London, South Africa on December 22. The curator of the local museum, Marjorie Courtenay-Latimer, was informed of its catch but was unable to identify it because no-one had ever seen one before. Eventually, in February 1939, its badly preserved remains were identified by Professor James L.B. Smith from Rhodes University, Grahamstown. He recognized it as a member of the lobed-finned fishes, the Crossopterygii, and called it 'Old Fourlegs' because of its lobed pelvic and pectoral fins, which he mistakenly thought enabled the fish to creep along the sea bed. He published a book on the subject: *Old Fourlegs*.

The fish was named *Latimeria chalumnae* after its discoverer (*Latimeria* after Miss Latimer; *chalumnae* after the river Chalumna, near the mouth of which the fish was caught). For nearly 14 years after its catch Professor Smith encouraged and organized searches for more specimens but the next recorded catch was not until 1952. Since the 1960s many specimens have been caught and by the 1980s scientists began to worry about the conservation of this rare species. Since 1991, coelacanths have been completely protected under the Convention of International Trade in Endangered Species (CITES).

Living specimens in their natural environment were filmed in the late 1980s by Hans Fricke using a deep-water submersible craft at 200 m, and millions of television viewers saw a dying specimen, filmed when Sir David Attenborough made his series *Life on Earth* in 1991. In 1997, a second species of *Latimeria* was discovered. *Latimeria menadoensis* lives in the Celebes Sea, between the Philippines and Borneo.

see also...

Living fossils

Coevolution

Evolutionary interaction between plants and animals has resulted in a phenomenon called coevolution. The term is usually reserved for pairs (or more) of species in which developments have taken place that confer benefits for both partners.

An association between birds and some plants is commonplace. Some plants produce fruit, which provide birds with food, and the fruit have seeds which are dispersed by the birds that eat them. The plants have evolved fruits with brilliant colours and nutritious juices to attract the birds that disperse the seeds. Pollination of plants by insects has involved changes in the structure of both partners. The shapes and colours of flowers, their scents and nectar, are all present purely to attract and reward the insects that pollinate them.

Coevolution is used by some authorities to mean any feature which has resulted from feeding relationships acting as selective pressures. In ecosystems there are close relationships between animals and plants and in the course of evolution organisms have developed mechanisms to protect themselves from competition. Many of these relationships have resulted in the production of protective toxins by plants and the development of counter measures, in the form of the ability to detoxify them, in hervibores. Studies on the relationships between different species of fruit flies, *Drosophila*, and the cactus, *Lophocereus schottii*, which produces toxic alkaloids, have proved the point. This cactus was found to be lethal to eight species of *Drosophila* but not to *D. pachea*, which, having evolved a detoxifying agent, was able to use the rotting stems of the cactus as a breeding place.

In Britain, the ability to produce highly toxic hydrogen cyanide (cyanogenesis) has evolved independently in many species of flowering plants, ferns and fungi. Cyanogenic and acyanogenic plants are eaten by the larvae of certain insects. These produce detoxifying chemicals which counteract the hydrogen cyanide.

Comparative anatomy

Students of comparative anatomy, or homology, have known for a long time that the species that make up any larger group (phylum, class or order) have numerous resemblances in structure. The development of techniques of genetic profiling have firmly established evolutionary links between members of particular groups by clearly showing similarities in their DNA. For example, all the species of the Phylum Chordata have, at one time in their lives, a notochord (the forerunner of a backbone), gill slits in the pharynx, and a dorsal hollow nerve cord. The species within a class would have more in common, and the species within an order still more. The species within a genus might be so much alike that only a specialist could tell them apart in terms of external features. Genetic profiling is particularly useful under these circumstances.

These similarities are interpreted as a consequence of evolution from a common ancestor. The classes within the Phylum Chordata are alike because, at one time in the past, they had the same ancestor. The resemblances of the members of a class (such as the Mammalia) would be still closer, because they would have had a common ancestor which had all the basic mammalian features.

Evolution tends to be a conservative process: rather than develop new structures, it tends to remodel existing ones. Thus, the fishes developed exceptionally complex structures that support the gills. In land vertebrates, these structures have become modified into other organs, such as the upper and lower jaws and bones of the middle ear. The muscles of the human face are very largely derived from the muscles of the gill arches of ancient fish.

Sometimes an organ is not remodelled into something new during evolution, but becomes reduced – and may even lose its function altogether. Such organs are called vestigial. Good examples are the tail bones, ear muscles, and appendix of humans. The whale has skeletal vestiges of hind limbs buried in the flesh where its tail begins. The python has tiny bony structures beneath the skin – these are all that remain of its hind legs.

see also...

Biogeographical evidence; Fossil evidence; Genetic profiling; Recapitulation evidence

Continental drift

In 1912 Alfred Wegener suggested that, about 200 million years ago, all of Earth's continents were joined into one enormous land mass, which he called Pangaea. In the ensuing 200 million years, Pangaea broke apart and the fragments began to drift to their present locations. The idea was not accepted until the 1960s, when data was obtained to provide proof of Wegener's theory. The most plausible evidence was based on magnetism in ancient lava flows. When a lava flow cools, metallic elements in the lava are oriented in a way that provides permanent evidence of the direction of the Earth's magnetic field at the time, recording for future geologists both its north–south orientation and its latitude. From such maps, it is possible to determine the ancient positions of today's continents. Continental drift is, of course, still continuing.

Geologists have long maintained that the Earth's surface is a moving crust, constantly changing, sinking and rising because of the massive forces beneath it. These constant changes are known to involve large, distinct segments of the crust known as tectonic plates. At certain edges of these plates, immense ridges are being thrust upwards, while other edges sink. Where the plates are crumpled together, the buckling has produced vast mountain ranges. When such ridges appear in the oceans, water is displaced and the oceans expand. With the use of astoundingly precise instruments, satellite studies have shown that Britain is moving away from the USA by 5 cm every year.

An understanding of continental drift (plate tectonics) is essential to the study of the distribution and evolution of life on our planet. It helps to explain the presence of tropical fossils in Antarctica and the unusual animal life of Australasia and South America. Continental drift provided just the sort of separation of populations that would permit widespread speciation, forming the basis for widely diverging groups of primitive organisms.

Pangaea began to split up in the Triassic (225 million years ago) and by the end of the Mesozoic era (65 million years ago) our present continents were recognizable.

Convergent evolution

onvergent evolution (convergence) is the development of similar structures in unrelated organisms as a result of living in similar ecological conditions. The development of the wings of vertebrates and insects is an example of convergence, in which quite distinct groups of animals have independently adapted to life in the air. In these animals the wings are said to be analogous structures, whereas wings of birds and bats have developed from the same pentadactyl ancestor and are therefore called homologous structures. When related animals develop the same feature in response to the same environment, such as the development of similar limbs in whales and seals, this is called parallel evolultion.

In ecosystems, particular ecological niches are occupied by individual organisms. To enable them to live in these particular environments, natural selection has imposed a particular shape upon each species. Obviously, a heavy animal with meat-shearing teeth could not live in a tree and eat nuts, fruits, and berries. A tree-living animal would ideally be small, so that the branches could take its weight. The animals would have long digits and claws, to grasp twigs, and accurate stereoscopic vision to help it to judge distance when it jumped.

One of the best illustrations of convergent evolution has been the adaptive radiation of Australian marsupials (pouch-bearing mammals) to occupy many niches that are occupied by placental mammals (all mammals except marsupials and egg-laying types) in the rest of the world. There are marsupial equivalents of anteaters, moles, shrews, cats and mice, with remarkable similarities in outward appearance to their placental counterparts.

see also...

Natural selection; Niche; Pentadactyl limb

Creationism

reationists vary in the way they interpret the story of creation, as written in the book of Genesis of the Bible or similar accounts in the fundamental texts of other religions. Some maintain that creation was completed in six 24-hour periods, and that all the living human races are descended from Noah. Others are prepared to accept that the *days* of creation in Genesis can be taken as geological eras, as long as God is accepted as the sole causative agent. Non-Christians tend to regard such attitudes as intellectually regressive, and indicative of the irrelevance of Christianity. Others are both convinced Christians and orthodox evolutionists. The views can be summarized thus:

★ Christians believe that their God is active through natural as well as supernatural acts. God is the creator and at the same time is accepted as working through mutation, selection etc.
★ Humanity is described in the Bible as emerging from a process similar to that of other organisms, but then *changed into God's image*. There is no reason to believe that the process of spiritual creation would alter us as a species physically and physiologically, although Christians believe that it makes us distinct from the rest of living things and gives a special relationship to God through a soul.
★ Humans, *made in God's image*, can fall from this state. This removes one of the Christians' main problems about evolution: the belief that acceptance of evolution automatically implies that humans are still on the way up, and hence improving.

The greatest amount of anti-Darwinian propaganda comes from creationists rather than scientists, although a new cult of creation-science has developed. The Evolution Protest Movement was founded as long ago as 1932 in Britain, with stated aims of publishing scientific information supporting the Bible and demonstrating that the theory of evolution is not in accordance with scientific fact. Its membership had grown to 850 by 1970, encouraged by the formation in the USA of the Creation Research Society in 1963 and the Institute for Creation Research in 1970. Further support for creationism in Britain was received by the formation of the Newton Scientific Association in 1972.

Cuvier, Baron C

Baron Georges Cuvier (1769–1832) dominated zoological thought in the early nineteenth century and made immense contributions to ideas concerning evolution because of his reputation as a comparative anatomist and palaeontologist. He was born in Montebeliard in the Jura and was destined to become the most powerful and influential scientist in France at the height of his career. After studying at the academy in Stuttgart, he became an instructor to the son of a rich family in Normandy. Here, his comprehensive studies on marine animals contributed to his book *The Animal Kingdom*, which was published in 1830. He became a friend of Tessier, a notable French scientist, who got him a job at the Jardin des Plantes in Paris. At the same time he also gave lectures on comparative anatomy at the university; these were recognized as new ideas in zoological thinking.

In 1795 Cuvier held various high positions in the newly built Natural History Museum in Paris. He later published a great work on fossil bones and was one of the first scientists to use fossil evidence to demonstrate that extinction had happened. Cuvier believed in creationism, to which he added the theory of catastrophism, believing that new life was created following each catastrophe. His studies on fossils quarried from the Paris Basin established the former existence of exotic animals. These fossils, and others including some gigantic ones that were sent to him from South America, effectively suggested the idea of extinction. He visited England in 1818 and examined many early collections of dinosaur fossils, but died before he could publish his observations.

One of Cuvier's most brilliant contributions was the identification of the first pterosaur ever to be found, in 1801. Nothing like it had ever been seen before but Cuvier applied his rigorous knowledge of comparative anatomy and recognized it as a flying reptile. He named it the pterodactyl (meaning 'wing finger') and, from the structure of its jaws and teeth, suggested that it fed on insects.

see also...

Creationism; Pterosaurs

In 1924, the Australian-born anatomist Raymond Dart found the fossilized skull of a child in a box full of pieces of limestone from a collection of fossils made by a South African mine manager of a quarry at Tuaung ('the place of the lion'), on the north-eastern edge of the Kalahari desert. It proved to be one of the most important discoveries of fossil evidence of human origins in Africa ever made. Dart was just about to leave his home to act as best man at a friend's wedding when he took a cursory glance at the box of limestone fragments and fossils that had been delivered to him during the day. He was amazed to see a fossil that he later claimed to be the 'missing link' between apes and humans and therefore the direct ancestor to ourselves. It was to become known as 'Dart's Child'.

At the time, he was written off as a heretic because the prevailing view was that the first step in the evolutionary divergence of humans from apes was the development of a large brain. The features of the Tuang child were the reverse: it had a small cranium (indicating a small brain), and human-like jaws and teeth.

Before Dart's discovery, Asia, not Africa, was considered to be the 'cradle of mankind' but Dart and Louis Leakey destroyed this myth.

The anthropological experts of the 1920s dismissed Dart's claim that the Tuang child, *Australopithecus africanus* ('African ape of the south') had anything to do with the origin of humans. Dejected and rejected, Dart virtually gave up research on the origin of hominids. No further fossils resembling the Tuang child ever came to light at the quarry. Indeed, in the 1990s, it was suggested that a prehistoric bird of prey carried it there because marks on the skull could be interpreted as being made by a beak or talons. It was left to Robert Broom in 1938 to prove the experts wrong and vindicate Dart's original claim.

see also...

Broom; Leakey

Darwin, Charles

The 'Theory of Evolution' is usually coupled with the name of Charles Darwin whose genius in the late 1850s gave the world the new idea of natural selection.

Darwin was born in Shrewsbury, England the son of a wealthy doctor. At a local public school he received the usual course of Latin and Greek verses with classical history and geography. This curriculum did not contain any natural history and so he satisfied his interest in this field by reading books on natural history and foreign travel. He performed chemical experiments in the garden shed and wandered the countryside, fishing, shooting, and collecting minerals, rocks and insects. His father thought his pursuits were disgraceful.

This supposedly idle boy was eventually sent to Edinburgh to study medicine. The sight of two operations performed with no anaesthetics revolted the 17-year-old Darwin and after two years, when it was apparent that he was totally unsuited to medical studies, he returned to Shrewsbury. His family now decided that he should study for the Clergy and in 1828 he became a student of theology at Cambridge. He proved to be far from studious and continued to annoy his father, who disapproved of the company he was keeping and his way of life.

On graduating with a modest BA pass degree in 1821, Darwin was prepared to take Holy Orders when fate intervened. On the advice of his professor and mentor, John Henslow, Darwin joined HMS *Beagle* on a voyage around the world, as an unpaid naturalist. This was the turning point in Darwin's life because it set an apparently unambitious man with little distinction on the way to show the world his genius and depth of reasoning. The observations and conclusions that he made on this voyage became the foundations of one of the most famous books ever written in any language. In 1859, Darwin published *Origin of species by means of natural selection, or the preservation of favoured races in the struggle of life* – more commonly known as *The Origin of Species*.

see also...

Darwinism; The voyage of the Beagle

Darwinism

Darwinism is the concept of Darwin's explanations of the mechanism of evolutionary change. It maintains that: In any varied population of organisms only the best adapted to that environment will tend to survive and reproduce. Individuals that are less well adapted will tend to perish without reproducing. Hence, unfavourable characteristics, possessed by the less well adapted individuals will tend to disappear from a species, and the favourable characteristics will become more common. Over time, the characteristics of a species will therefore change, eventually resulting in the formation of new species.

The main weakness of Darwin's theory was that he could not explain how the variation, which natural selection acts upon, is generated. At that time it was believed that characteristics of the parents become blended in all offspring. This weakness was overcome with the discovery of Gregor Mendel's work and its description of particulate inheritance.

Another weakness of Darwinism was the fact that Darwin was not able to prove his hypotheses experimentally. However, he *was* familiar with the principle of artificial selection, although the intricacies of the genetic explanation were not revealed until the work of Mendel was recognized in 1900. For thousands of years it has been known that, through selection of animals and plants for breeding, certain desired traits could be passed from one generation to the next. Darwin demonstrated how artificial selection could be applied to pigeons and he bred many varieties at his home in Kent. In fact, this was probably the only experimental evidence that he had to reinforce his theory of natural selection. He deduced that nature could take the place of humans in the selection process and could determine the reproductive success of individuals. Darwin suggested that natural selection would be a more 'hit-or-miss' process than selection by humans.

see also...

Artificial selection in animals;
Artificial selection in plants;
Darwin; Darwin's finches;
Mendelism; Natural selection;
Natural selection by predation

Darwin's finches

In 1835, Charles Darwin visited the Galapagos Islands during the epic voyage of HMS *Beagle*. The finches that he found there are now named after him because they are famous for being one of the main inspirations for his theory of evolution by natural selection. The ancestors of these finches had presumably been blown to the islands by high winds and, once established, were able to evolve into several distinct forms to take advantage of unfulfilled niches available on the islands. The divergence from these ancestral finches to the several modern forms can be seen in many aspects of their lives.

They have become Darwin's finches, being distinct species of the Galapagos Islands. Apart from one species on the Cocos Islands, some 600 miles to the north-west, they are found nowhere else in the world. Their ancestors were from the South American mainland, 600 miles east.

In appearance, Darwin's finches are very drab and unprepossessing – greyish-brown with occasional patches of black on some males, and of the 14 species (of which Darwin discovered 13) there is little variation except in the beaks and diet. This is unusual, because on the mainland, finches usually differ in plumage and not in body form.

The shapes of the beaks are related to the type of food that the finches eat. They range from slender to stout and are responsible for the survival of the finches in their respective niches. There are few other species of birds on the Galapagos, so the immigrant ancestors of the finches found little competition. Where there was a possibility of competition, for example on the sea shore, no finches have succeeded in exploiting this particular niche. Probably because of competition from birds of prey and scavengers, none have become flesh-eaters, except that one species takes blood from nesting boobies and feeds on eggs. The wood-pecker's niche has been taken over by a finch that uses a cactus spine for winkling out grubs from holes in trees.

---

see also...

The Galapagos Islands; Natural selection; Niche; The voyage of the Beagle

Dinosaurs

During the Triassic period (245–202 million years ago), the reptiles called archosaurs (the 'ruling reptiles') evolved into four main groups: the two orders of dinosaurs; the pterosaurs, and the crocodiles. The two dinosaur groups ('Bird-hipped' and 'Lizard-hipped') were no more closely related to each other than they were to the crocodiles or the pterosaurs. So the term 'dinosaur', although popularly used, has no real scientific meaning as a single taxonomic group. The word, dinosaur, comes from the Greek words *deinos* (meaning terrible) and *sauros* (meaning lizard). It was coined in 1842 by Sir Richard Owen, the Director of the Natural History Museum in London.

Our knowledge of the two groups of dinosaurs is limited to that which can be gained by forensic means from their fossils. We know more about the types that lived in swamps or beside rivers, because carcasses that lay in initially wet lands were more likely to become fossils. We know much less about those that lived inland or in ancient mountainous regions. Those that we *do* know about are the four-legged bird-hipped dinosaurs with horns and armoured frilled necks. Their heavy skulls have been found in river and swamp sediments where they were washed down by mountain rivers and streams. In contrast, the skulls of swamp-dwelling dinosaurs are badly preserved because they are normally lightly built and easily crushed.

Probably the earliest written record of a fossil dinosaur is a Chinese description of a 'dragon' bone, made over 1700 years ago, and early American Indians used fossil bones and teeth as good luck charms. The first dinosaur ever to be scientifically described in Britain was *Magalosaurus* in 1824 by Dr William Buckland, Professor of Geology at Oxford University. It was a creature 7 m long with three huge claws on its hind feet. Its skull was equipped with serrated meat-shearing teeth and its head may have been held up on an S-shaped curve of its short neck, like that of a bird. By the end of the 1920s, fossil dinosaurs had been found on every continent.

see also...

*Bird-hipped dinosaurs;
Lizard-hipped dinosaurs; Owen*

Directional evolution

One objection that has been made to Darwin's theory of evolution was that the concept of change attributed to variation and natural selection could not explain the trend in some organisms to evolve in particular directions, hinting that their evolution was somehow directed towards a particular target.

The idea of a directional goal in the development of life is not new. Part of Lamarck's theory of use and disuse had been that all organisms possess an innate urge to increase in complexity, an idea that Darwin rejected when he wrote the first edition of *The Origin of Species*. On a broader plane was the assumption that the evolution of certain animals, such as the horse towards increasing height and gradual reduction in the number of digits, was somehow predetermined. The tendency for related groups of organisms to evolve in the same direction, irrespective of natural selection, became known as orthogenesis. As this theory implies that evolution proceeds independently of natural laws, it is impossible to test experimentally. Orthogenesis therefore conflicts with conventional evolutionary theory such as neo-Darwinism.

However, the idea of orthogenesis has proved particularly attractive to creationists, who have seen in it the control of the Creator and the manifestation of a Divine plan. One of the reasons why orthogenesis was popular among the post-Darwin Victorians was the paucity of fossil evidence for evolution at that time. We now know that nearly all instances of apparently linear evolution, including that of the horse, have been associated with much branching and adaptive radiation into a variety of niches. Many of these side branches soon became extinct but some survived for considerable periods in various parts of the world. Once a particular variation has become established, the chances that further development will take place in the same direction are greatly increased.

see also...

Darwinism; Lamarck;
Lamarckism; Natural selection;
Neo-Darwinism

Dubois, Eugene

In 1891, Eugene Dubois (1858–1940) found fossils in Java of an early hominid, *Homo erectus* (Java Man – literally, 'upright man'), now known to be 1.8 million years old. In 1877, aged only 19, Dubois began to specialize in anatomy and natural history at the University of Amsterdam. A few years later, he became obsessed with the idea of finding a really primitive ape-like fossil human and concluded that the East Indies might be the place to find it. His decision to search there was probably influenced by a role model of his, Ernst Haeckel, the prominent German biologist of the day. Dubois approached the Netherlands government to sponsor his expedition to Dutch territories in the Far East. Not surprisingly, the prospect of financing a professor's assistant to find a purely imaginary creature was not very high in the list of the government's priorities and their answer was brief – 'No'. However, the undaunted Dubois enlisted as a medical officer in the Dutch Army. At the age of 29, after resigning from his secure job as a lecturer in anatomy in the University of Amsterdam, he travelled to the Dutch East Indies with his wife and child.

There he was to make one of the greatest discoveries in the history of humanity – he was to find his 'missing link'.

Labourers from Dubois' team made the actual find and Dubois was not present to see the precious fossil brought to light. It was part of a skull which seemed to Dubois to be too large to be the cranium of an ape and too small to be that of a human. In May 1892 a femur was found about 13 m away from the first find and was thought to be part of the same individual. His diagnosis of an upright human walker with a primitive cranium is now reason to give the 'people-ape' the name *Homo erectus*.

However, on his return to his home country, Dubois was ridiculed for suggesting that he had found the 'missing link'. He was not vindicated until the 1930s.

see also...

Haeckel

Ediacara fauna

This group of fossilized animals is named after the Ediacara Hills of the Flinders range in South Australia, where they were first discovered. The Australian geologist R.C. Sprigg was responsible for finding this varied assortment of invertebrate fossils in 1947. Martin Glaessner is considered to be the first palaeontologist to describe them in detail and interpret them as primitive members of the group to which jellyfish and corals belong, but including some annelid worms and arthropods. These creatures dominated the Earth for tens of millions of years, originating about 680 million years ago in rocks which just pre-dated the Cambrian 'explosion' and the Age of invertebrates.

It took some extraordinary geological circumstances to make it possible for these organisms to leave any remains at all because they were all soft bodied, without skeletons. The Ediacara Hills provided the right conditions at the right time, their fine sandstone sediments allowing the outlines of soft jelly-like bodies to be preserved. Proving that the Ediacara fauna were not just a 'one-off freak', subsequent discoveries have shown that their distribution is world-wide.

According to the dating of some fossils found in Namibia in the late 1990s, they lasted until the Cambrian period (540–505 million years ago). They were related to modern-day invertebrates but their external features do not resemble any survivor very closely. Indeed, some scientists believe that the Ediacarans were neither animal nor plant but showed features of both groups.

In the 1980s, Dolf Seilacher, Professor of Palaeontology at Tubingen, Germany, claimed that the Ediacara fauna might have been a failed evolutionary experiment in multicellular organization because no Ediacara elements survived into the Cambrian. It seems that the development and evolution of the Ediacara animals ended before the beginnings of the more advanced internal organ systems that are found in higher invertebrates and recognized in the Cambrian fauna. The alternative pathway of evolving large surface areas without internal organs seems to have resulted in very few groups that could be thought of as being successful.

see also...

The Age of invertebrates

Eukaryotes

ukaryotes can be single-celled organisms or organisms made of many cells whose genetic material, in the form of DNA, is enclosed by membranes (the nuclear envelope) to form a nucleus. This is in contrast to a prokaryote, which is more primitive in that there is no nuclear membrane and fewer intracellular structures (organelles). Eukaryotic cells are like factories, with the nucleus representing the management, the other organelles the machinery, and the enzymes the workers. Like a factory, the cell must obtain raw materials for it to perform efficiently. It must also be able to get rid of waste.

In all cells, the cell membrane is composed largely of lipid (fat) molecules called phospholipids, together with protein molecules. The formation of these biological molecules – lipids and proteins – was therefore a prerequisite for the development of the membranes around the cells of all living organisms. Each phospholipid molecule is made of a water-soluble 'head' and a water-repellent 'tail'. The molecules line up in two rows: the heads being attached to the watery cytoplasm, point outwards; the tails, being repelled by the water, remain hidden inside the membrane. This arrangement is both stable and flexible. It is stable because the opposition between the phospholipid water-soluble heads and water-repellent tails prevents the molecules turning around so that their tails, not being rigid, wave around inside the membrane.

Besides being essential selective barriers to the external environment of cells, membranes are also important within cells. They divide the cell into compartments called organelles, where different functions can be carried out and many enzyme-controlled reactions take place. Internal membranes have the characteristic structure of the outer membrane, suggesting that they might have evolved from folds of the outer layer. A double membrane maintains the integrity of the control centre of the eukaryotic cell – the nucleus.

see also...

Prokaryotes; Protobionts

Extinctions

The most famous of all the mass extinctions which have taken place to date was the one at the end of the Cretaceous period, about 65 million years ago. The cause is thought to have been a collision with a gigantic comet, but when this suggestion was first put forward in the 1980s many scientists were very sceptical.

It was Luis Alvarez and his son, the geologist Walter Alvarez, who provided the evidence which made the explanation irrefutable. They had found massive levels of iridium in the thin layer of clay that marks the exact boundary between the Cretacous period and the Tertiary era. They first found this layer at Gubbio in central Italy, and similar deposits were subsequently found in Mexico, Antarctica and Tunisia. Indeed, all over the world, where scientists have examined the Cretacious–Tertiary boundary layer, they have found iridium. The element is found in deep mantle rocks but is very rarely found on the surface of Earth. It has been observed that iridium can sometimes reach the Earth's surface as a result of volcanic activity but comets and meteorites are also rich sources.

The impact site has been established as the Chicxulub crater in the Yucatan Peninsula, Mexico. Such a crater can be explained by the impact of an asteroid about 10 km across. In 1996 evidence was provided to show that the fateful asteroid hit the peninsula from an angle of 20–30° above the southern horizon, showering molten debris towards the north-west. There would have been a massive cloud of vaporized and molten rock, saturing the atmosphere over the west of North America and instantly roasting any living thing in its path. Fragments ejected by the impact would have re-entered the atmosphere like smaller meteorites, with further devastating consequences. The blast would have stripped some of the Earth's atmosphere into space. Computer simulations have estimated winds of at least 500 kph and that all combustible material would have produced 70 billion tonnes of soot – the equivalent of burning 25 per cent of all organic matter on Earth.

see also...

Catastrophes

Fossils

Fossils are the remains of, or impressions left in rocks by, extinct organisms. Most fossils consist of hard skeletal material because soft tissues and organs of dead organisms are either eaten or decay rapidly. Mineral salts from surrounding rocks gradually replace the hard organic material, to give a cast in a process called petrification. Alternatively, the organic material dissolves, leaving an impression or mould in the surrounding rocks. Trace fossils, such as the footprints of dinosaurs or tubes and burrows of worms, often provide evidence of the former existence of extinct forms. Fossilized faeces, called coprolites, also come under this category.

Occasionally organisms are preserved in other ways, as in the many examples of insects trapped in ancient tree resin, preserved as amber. Deep-frozen mammoth remains are frequently discovered in Siberia, some of which are so complete that their stomach contents are still recognizable. When some of these were first discovered by explorers using dog sledges, the dogs were able to eat the meat of these extinct animals.

The most important conclusions from the fossil record:

★ The simplest forms of life appeared first and survived, in some cases for hundreds of millions of years before the evolution of more advanced forms.
★ When fossils are arranged in chronological order, clear trends of evolutionary descent often become obvious. Changes in structure can be interpreted as adaptations to new conditions as a result of selective pressures acting on a former design.
★ Extinction is a frequent event so that only a very few forms are represented by the descendents of extinct forms, today.

There are many instances of fairly complete fossil records showing the progressive evolution of certain types of organisms. Ammonites and some sea urchins are good examples. In these cases intermediate stages are common.

see also...

Fossil evidence; Punctuated equilibria

Fossil evidence

The hard evidence for evolution has been recorded by the thousands of fossil animals and plants that palaeontologists and geologists have dug out of rocks. The record tells a story of trends in the development of life on Earth for thousands of millions of years. It also shows how Earth itself has changed. For instance, the deserts of Wyoming, now more than a mile above sea level, contain fossils of hundreds of types of fish that once lived in the sea that has vanished from the area. Further south, in Arizona, a desert can be found containing fossilized trees that made up a vast forest millions of years ago. This is evidence that massive climate change has taken place, from moist, mild conditions to the harsh desert that we see today.

Fossils have been formed and preserved by a succession of coincidences because dead organisms are usually eaten or decay. The chances of discovering a fossil are quite remote and the chances of finding a series of fossils which trace the evolutionary ancestry of an organism even remoter. However, examples of almost complete lineages of fossil ancestors have been found.

Usually they are fossils of shelled animals or of creatures with very robust skeletons or teeth. For example, a small four-toed animal about the size of a cat was found fossilized in sedimentary rocks some 60 million years old. From this animal can be traced a gradual line of descent, with continuing modifications, right down to modern horses, with their relative fossils being found in progressively younger and younger strata of rocks. Similar evolutionary sequences have been established for the elephant, the giraffe and the camel. Even more detailed changes through time have been traced for some invertebrates, including certain molluscs and sea urchins.

Fossil evidence dramatically shows that life has been gradually changing over millions of years from one form to another. There is no longer any reasonable doubt that evolution occurs, and that it has occurred since the origin of life.

see also...

Biogeographical evidence;
Comparative anatomy;
Recapitulation evidence

Founder effect

he principle of the founder effect was put forward as an explanation of the evolution of small populations by Ernst Mayr. Sometimes a few organisms will be separated from the main population and establish their own breeding group. For example, a few birds might be blown from the mainland to some remote island, as probably happened in the case of Darwin's finches. These became the founders of a new population. Sometimes storms might wrench a large tree loose, allowing it to float out to sea. Such a raft could carry small creatures and eventually strand them on new land, perhaps an island. Humans have unwittingly introduced rats and other small animals to islands in this way, with disastrous consequences.

Evidence of natural rafts for transporting animals to islands was obtained in 1995 when Ellen Censky of the Carnegie Museum of Natural History in Pennsylvania recorded that 15 iguanas drifted on a mat of vegetation to Anguilla from Guadeloupe, a journey of about 300 km. This followed Hurricane Louis, when hillside was washed into the sea. The iguanas arrived in Anguilla alive, and one healthy female was sighted there 29 months later.

The first colonists of an island, or the first individuals to spill over into a previously unoccupied area, are the founders of a new population. If relatively few individuals are involved, they may carry an unrepresentative selection of genes from the parent population. When this happens, the new population, descended from the founders, will have a divergent gene pool from the outset. Natural selection may act to increase divergence, producing a distinct group of organisms in the new region.

An example is the meadow brown butterfly, *Maniola jurtina*. On the island of Tresco in the Isles of Scilly, south-west of Cornwall, England, a population of this butterfly had a distinct pattern of spots on its wings that was quite different from the pattern of the populations found on the mainland. Within an accurately measured time a new population was evolving from a founder population of less than 200 individuals.

see also...

Darwin's finches; Genetic drift; Genetic profiling; Natural selection

Galapagos Islands, The

Charles Darwin's visit to the Galapagos Islands in 1835 proved to be the most important part of the voyage of HMS *Beagle* when he came to propose his theory of natural selection. The Galapagos Islands are named after the Spanish for tortoise, *galapago*, but they are also called the Enchanted Isles. They comprise a chain of volcanic islands, 580 miles off the coast of Ecuador and form unique ecosystems of animals that have become specialized due to their isolation. Darwin wrote of them: 'A little world in itself with inhabitants such as are found nowhere else'. There were marine iguanas, half a metre long, grazing on seaweed beneath the sea, which Darwin called 'imps of darkness, black as porous rocks over which they crawl'. He became aware of giant tortoises that differed in shapes of their shells and in their modes of feeding, depending on the island they occupied. These giant tortoises no longer live on the mainland, although their fossils exist there. On the Galapagos Islands today, there are ten distinct races of tortoise (originally there were at least 15 races but over the past 400 years, five have become extinct through human activities). There is only one survivor of a particular race that used to live on the island of Pinta. This last specimen is housed at the Charles Darwin Research Station on Santa Cruz.

In 1957, a group of scientists working with international conservation groups and *Life* magazine recommended that some of the islands should be set apart as sanctuaries to protect the unique fauna and flora. The Charles Darwin Memorial Station was set up as a result and is now an international study centre.

Perhaps the most well known of Darwin's observations concerning the islands were the result of his studies on the islands' bird life – particularly the finches. Darwin noted 13 species, all of which had evolved from an original ancestral stock from the mainland. He suggested that the birds colonized the islands largely as a result of their adaptations which enabled them to exploit available food sources.

see also...

Darwinism; Darwin's finches; Natural selection; The voyage of the Beagle

Genetic drift

This is sometimes called the Sewall Wright effect after the two population geneticists who suggested it. Once Gregor Mendel had shown how genes are inherited and Hardy and Weinberg had demonstrated how these genes are expected to behave in populations, biologists realized that evolution might occur by chance as well as being directed by natural selection. Genetic drift depends on the fact that the fluctuation of allele frequencies in a small population is due entirely to chance. If the number of matings is small then the actual numbers of different types of pairing may depart significantly from the number expected on a purely random basis. Genetic drift is one of the factors that can disturb the Hardy–Weinburg equilibrium.

Randomly mating populations are profoundly influenced by the powerful effects of natural selection. In these groups, as adaptive traits are selected, others are relentlessly winnowed out as the populations become adapted to the environment by directional selection. Small populations behave differently and are modified by different factors than those operating in larger populations. For example, in small populations there is a change in frequency of genes in a population due to simple chance. Changes of this type are not produced by natural selection. Genetic drift is important to small populations because they would, by definition, have a smaller gene pool (reservoir of genes). This means that any random appearance or disappearance of a gene would have a relatively large impact on gene frequencies. In large populations, any random change would have little impact because it would be swamped by the sheer numbers of other genes in the population. The bottle-neck effect is an important way that small populations can set the stage for such random events.

By definition then, genetic drift is the random fluctuation in gene frequencies due to sampling error. In nature, it is important in small populations that become isolated, such as island colonisers, koala bears or the giant panda.

see also...

Bottle-neck effect;
Hardy–Weinberg equilibrium;
Mendelism; Natural selection

Genetic profiling

In the 1980s Professor Alec Jeffreys at Leicester University showed the presence of many variable regions of DNA which did not code for amino acids. These regions were called minisatellites. There are thousands of these scattered throughout chromosomes, probably having evolved as mistakes during replication of DNA. The number of times that these regions are repeated gives individuality to the profile. If enough regions of DNA are examined, it is possible to obtain a genetic profile which is almost unique to an individual.

The chance of two people having the same genetic profile is one in a million unless they are identical twins.

A genetic profile is made as follows:

1 Cells with nuclei are obtained from any tissue.
2 DNA is extracted from the nuclei.
3 The DNA is cut into fragments using restriction enzymes.
4 The fragments are separated according to size using electrophoresis. This is the process of subjecting the mixture, on a gel soaked in a conducting solution, to an electric field. Electrically charged DNA fragments move towards oppositely charged electrodes: the positive move to the cathode and the negative to the anode. The rate of movement varies with the size of the fragment.
5 The fragments are transferred to a nylon membrane in a process called Southern blotting. The nylon membrane is sandwiched between the gel and sheets of blotting paper. The DNA fragments move into the membrane.
6 Sodium hydroxide is added to the membrane. It splits the DNA into single strands, leaving the sequence of bases intact.
7 The DNA fragments are identified with a DNA probe. This is a portion of DNA with a base sequence that is complementary to that of the minisatellite. The probe is labelled with a radioactive tracer which affects X-ray film.
8 X-ray film is placed over the membrane and developed. The result is a pattern of bands something like a shop barcode, showing where the probe has bound to the DNA.

Geographical isolation

Geographical barriers have had a major impact on the development of new species. If populations cannot reach each other, they cannot interbreed. Continental drift, mountain formation, and volcanic action have all taken part in providing barriers to separate populations within a species. The remnants of larger bodies of fresh water, in the formation of isolated lakes, has led to the evolution of separate subspecies of fish. An example is the rainbow trout in the USA. They evolved from a single species which once was able to reach all of the, now isolated, lakes and rivers. In Britain, the char, a relative of the salmon, remains in some isolated lakes as a result of changes in the last Ice Age glaciation.

The presence in Australia of marsupial mammals but no ancient placental mammals is also due to physical change – that of the position of the continent itself. The marsupials originated in North America, from where they spread to South America and Europe. They became extinct in Europe and North America because they were unable to compete with the placentals there.

From South America, however, they moved across what was to become Antarctica and into the area that finally became the island continent of Australia. In the absence of competition from placentals, they were able to evolve into the great variety of forms that we see today by divergent evolution and adaptive radiation.

Islands arise in a variety of ways but whether they are formed by coastal erosion, volcanic action or as coral formations, they all have a profound influence on the evolution of organisms that live on them. Natural selection results in rapid change and diversity of immigrants to the islands.

On long-established islands, natural selection has resulted in forms that are markedly different from the ancestral stock that arrived there. Darwin's finches are good examples.

see also...

Adaptive radiation; Allopatric speciation; Continental drift; Darwin's finches; The Galapagos Islands; Natural selection; Niche; Reproductive isolation

Geological time

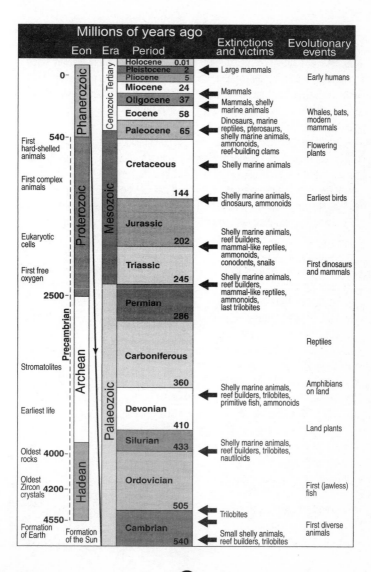

Millions of years ago				Extinctions and victims	Evolutionary events
	Eon	Era	Period		
0–	Phanerozoic	Cenozoic Tertiary	Holocene 0.01	Large mammals	Early humans
			Pleistocene 2		
			Pliocene 5		
			Miocene 24	Mammals	
			Oligocene 37	Mammals, shelly marine animals	
			Eocene 58		Whales, bats, modern mammals
			Paleocene 65	Dinosaurs, marine reptiles, pterosaurs, shelly marine animals, ammonoids, reef-building clams	
First 540– hard-shelled animals		Mesozoic	Cretaceous	Shelly marine animals	Flowering plants
First complex animals	Proterozoic		144	Shelly marine animals, dinosaurs, ammonoids	Earliest birds
			Jurassic	Shelly marine animals, reef builders, mammal-like reptiles, ammonoids, conodonts, snails	
Eukaryotic cells			202		First dinosaurs and mammals
First free oxygen			Triassic 245	Shelly marine animals, reef builders, mammal-like reptiles, ammonoids, last trilobites	
2500–			Permian 286		
	Precambrian Archean	Palaeozoic	Carboniferous		Reptiles
Stromatolites			360	Shelly marine animals, reef builders, trilobites, primitive fish, ammonoids	Amphibians on land
Earliest life			Devonian 410		Land plants
Oldest 4000– rocks			Silurian 433	Shelly marine animals, reef builders, trilobites, nautiloids	
Oldest Zircon 4200– crystals	Hadean		Ordovician		First (jawless) fish
4550– Formation of Earth		Formation of the Sun	505	Trilobites	First diverse animals
			Cambrian 540	Small shelly animals, reef builders, trilobites	

47

ɟinkgo biloba

An evolutionary breakthrough came with the development of a group of plants called the cycads. These resemble palm trees, with large leafless stems and stiff leaves. About 120 million years ago they were widespread but today they are restricted to tropical and subtropical regions of the world. They were some of the first seed-producing plants. A close living relative of these ancient cycads has a good claim to being the most senior surviving plant in the world and can therefore be classed as a 'living fossil'. It is the maidenhair tree, *Ginkgo biloba*, the sole remaining representative of the Ginkgoales, being of the same generation as the dinosaurs. Its unique, fan-shaped, bilobed leaves have been fossilized in Permian rocks (286–245 million years old) and it is possible that they appeared in the Carboniferous period (236–286 million years ago).

Ginkgo biloba survived in Britain until the middle Tertiary (about 50 million years ago). During the Jurassic period (about 200 million years ago) it reached its greatest abundance and had an almost world-wide distribution. Its disperal was probably aided by the prolonged resistance of its seeds to immersion in sea water. It has declined progressively since that time and it is most unlikely that it exists in its truly wild state anywhere in the world.

The famous maidenhair tree of Kew Gardens in London was planted in 1760. However, in China and Japan, it has been grown from time immemorial as a temple tree. Indeed, it was probably saved from extinction only by its adoption by the ancient Chinese as a sacred plant. Today it is found in many cities of the world in parks or as a deciduous 'street tree' because it is highly resistant to insects, disease and air pollution.

The name *Ginkgo* is something of an etymological enigma. It is thought to be an erroneous copy of 'Gin-ko', which is the Japanese equivalent of two Chinese symbols, meaning 'silver fruit', a reference to its plum-like fruits. The seeds from its fruits are still sold in Chinese markets as roasted *sal* nuts.

see also...
Living fossils

Gradualism

The concept of whether evolution proceeds by gradualism or that it is episodic by punctuated equilibria stems from discussions and research by biologists which go back to the early 1940s. At that time Ernst Mayr suggested that some present-day genera of birds had evolved rapidly, each as a small marginal population of the ancestral species. In such small and localized populations evolutionary change would be rapid and so, if a similar process had occurred when a fossil series showed that branching had happened, it would be very unlikely that fossils of the actual diversification would ever be found. Mayr's views were strongly supported in the 1970s by Niles Eldredge and Stephen Jay Gould, who first gave the terms 'punctuated equilibria' and 'gradualism' to these alternative views on how evolution might have taken place.

The evolutionists of the day became divided in their views as soon as current thinking was challenged. There were the strong punctualists on one hand and equally strong gradualists on the other. The debate has continued ever since and has led to an assumption by many that the two sides are arguing about whether evolution does or does not take place. What is *not* always appreciated is that both sides accept that present-day species have evolved from other species, and that the debate is really about the possible mechanisms that could have been responsible for the process which is still continuing.

The sudden appearance of new species in the fossil record can, of course, be taken by some creationists, who do not restrict themselves to the biblical account, to be evidence of periodic acts of special creation by a supernatural being, after catastrophes made previous ones extinct. Such views are untestable.

Punctualism and gradualism are important concepts in explaining how new species evolve. A major problem of comparing the two viewpoints is that fossil evidence is confined to hard, skeletal remains. Nothing is really known about the lineages of soft parts, behavioural patterns or the ecological conditions in which extinct organisms evolved.

see also...

Creationism; Punctuated equilibria

Haeckel, Ernst Heinrich

Ernst Heinrich Haeckel (1834–1919) was a German naturalist, well known for his early adoption of Darwinism and for his construction of geneaological tree of living organisms. He was born in Potsdam and in 1862 became Professor of Comparative Anatomy and Director of the Zoological Institute at Jena. He spent most of his life at Jena except for the time he devoted to various lecture tours. Haeckel was a field naturalist and the person who first coined the word 'ecology', in 1866.

He lived at the same time when there was much controversy over Darwin's *Origin of Species*, which was published in 1859, but, having accepted Darwin's doctrine, he dedicated much time to popularizing Darwinism throughout Germany. The major work for which Haeckel is reknowned is his Theory of Recapitulation. Von Baer first put forward theories on the development of individuals (ontogeny) in 1828, and these were followed by even more detailed research by Johann Muller in 1863. It was Haeckel, however, who imposed an evolutionary interpretation on the embryological laws and who expanded the original idea, so they became mistakenly ascribed to him in 1866.

The recapitulation theory suggests that an embryo of any species undergoes in its development, the evolutionary or phylogenetic history of its race. In a nutshell – 'ontogeny repeats phylogeny'.

In his original publication, von Baer said that the young stages in the development of an animal are not like the adult stages of other animals lower down the ancestral scale but correspond to the young stages of those other animals. Haeckel's theory, however, stated that characters of the adult ancestor, not the young, are repeated in the embryo of the descendant. An example would be that the gill pouches of an embryonic bird repeat the gill slits of its adult fish ancestors. The recapitulation theory is considered to be fundamentally flawed because the gill pouches of an embryonic bird resemble the embryonic gill slits of fish ancestors.

see also...

Darwinism; Recapitulation evidence

The Hardy–Weinberg equilibrium

The concept stems from the Hardy–Weinberg principle. It shows that the frequency of alleles for any character will remain unchanged in a population through any number of generations unless it is altered by some outside influence.

Imagine that **A** is dominant to **a**. A cross between **AA** and **aa** will give an F_1 of **Aa**. If we calculate the gene frequencies in the F_2 generation from a cross between the F_1, we see that a quarter of the population is **AA**, half is **Aa**, and a quarter is **aa.** So, to determine what fraction of the F_3 generation will be offspring of, say, **AA** and **Aa** crosses, we simply multiply one-quarter by one-half and get one-eighth. In the third generation then, assuming random mating, we can expect one-eighth of the population to have genotypes resulting from this cross. What part of this generation will be the offspring of **AA** and **AA** crosses? Of **Aa** and **aa** crosses? Each of these combinations can be expected to produce one-eighth of the F_3 generation, so what part of the population will be heterozygous (**Aa**)? By counting the frequency of this type of gamete among the possible combinations, we see that $4 \times$ one-eighth = one-half. And what

was the fraction of **Aa** individuals in the F_2 population? One-half! If you work out the results for all possible combinations in the F_3 generation, or for any combination in any generation after the F_3 generation, you will see that the ratio of genetic components in a population remains stable. This is why we continue to have blue eyes in our population, even though they are a recessive trait.

The conditions required for the Hardy–Weinberg equilibrium are as follows:

★ Large population.
★ Random mating. Every individual of reproductive age has an equal chance of finding a mate.
★ No emigration or immigration (no gene flow).
★ No selection pressure (no natural selection).
★ No mutation.

Any of the above can occur in natural populations, so the Hardy–Weinberg model is to a great extent an artificial one.

see also...

The Hardy–Weinberg principle

The Hardy–Weinberg principle

In 1908 a lunchtime discussion took place at Cambridge University between the geneticist Reginald Crundell Punnett and his older friend G.H. Hardy. Punnett said that he had heared an argument critical of Mendelian philosophy that he couldn't answer. According to the argument, if a gene for short fingers were dominant, and the gene for long fingers were recessive, then short fingers ought to become more common over each generation. Within a few generations, the critics said, no one in Britain should have long fingers!

Punnett disagreed with the conclusion but couldn't give a good reason. Hardy said he thought the answer was simple enough and jotted down a few equations on a table napkin. He showed the amazed Punnett that, given any particular frequency of genes for normal or short fingers in a population, the relative number of people with short or long fingers ought to stay the same as long as the population was not subject to natural selection or other outside influences that could lead to changes in gene frequency. (Gene frequency refers to the ratio of different kinds of genes in a population.) Hardy thought the idea too trivial to publish but Punnett talked him into publishing it somewhere other than on a table napkin!

Others were developing the same idea, including a German physician named Wilhelm Weinberg, so the idea came to be known as the Hardy–Weinberg principle. The principle was actually developed even earlier by the American, W.E. Castle, but strangely he is never mentioned in standard text books.

The principle is stated today as:

In the absence of forces that change gene ratios in populations, when random mating is permitted, the frequencies of each allele (as found in the second generation) will tend to remain constant through the following generations.

This led to the concept of the Hardy–Weinberg equilibrium.

see also...

The Hardy–Weinberg equilibrium

Heavy metal tolerance

Selection for heavy metal tolerance in some plants is an example of an environmental selective pressure influencing the evolution of populations. A number of grasses show genetic adaptation to high concentrations of metal ions in areas where spoil heaps from mines have brought potentially toxic heavy metals to the surface. In Britain, many deposits of waste from zinc and lead mining are well over 100 years old but are still devoid of plants; nevertheless, some species of grasses, *Agrostis* and *Festuca*, can colonize them. Plants from such sites show tolerance of the metal when grown in experimental conditions and since the 1970s more than 21 species of flowering plants have been identified as being heavy metal tolerant throughout the world. The tolerance is inherited and controlled by several genes.

The evolution of these strains of grasses has allowed them to colonize newly available niches. Tolerant individuals of *Agrostis* occur at a frequency of about one or two per thousand in populations not exposed to toxic metals and there is evidence to suggest that the successful growth of tolerant plants on non-toxic soils is adversely affected by competition with non-tolerant plants, so that the former tend to be eliminated. The disadvantage may not be very great, however, so that tolerant individuals persist in populations. The reverse is certainly not true on the spoil heaps. In such situations, non-tolerant plants are non-viable and only tolerant plants survive. The result is that patches of metal-tolerant plants may be of very small size, and they show a sharp boundary at the edge of toxic sites. Some zinc-tolerant individuals have been found in belts no more than 30 cm wide beneath zinc-coated metal fences. When a new toxic spoil heap is created containing high levels of zinc, copper or lead, rapid colonization by certain species of plants can be expected, starting from tolerant individuals present in normal populations. As selection for tolerance can be extremely strong, a population can change from almost entirely non-tolerant individuals to only the few tolerant survivors within a single growing season.

see also...

Niche; Selection

Ichthyosaurs

Perhaps the most well known of the marine reptiles of the Triassic and Jurassic periods (245–144 million years ago) were the ichthyosaurs. Superficially, they looked like modern dolphins, with a fish-like shape and a shark-like dorsal fin. Their limbs were adapted as balancing and steering paddles and they had a large propulsive fluke on their tails. Several genera of ichthyosaurs can be traced from the long and narrow *Mixosaurus* of the Triassic, through the dolphin-like *Ichthyosaurs* and the swordfish-like *Eurhinosaurus* of the Jurassic to the toothless *Ophthalmosaurus*. They were all active hunters and ate fish and the ancestors of squid.

The ichthyosaur paddle was a specialized organ, perfectly shaped for travelling through water at speed. It was strengthened by a mass of finger bones that seem to bear no resemblance to the arrangement of bones in a normal reptilian limb. This factor suggests that ichthyosaurs attained their marine form directly from an ancestral aquatic amphibian, without passing through a land-living reptilian stage. Their adaptation to marine life was so perfect that they bore their young live, rather than coming to land to lay eggs. Their eventual extinction at the end of the Mesozoic era (about 65 million years ago) was partly due to competition from other newly evolved marine reptiles.

Evidence of the former existence of these marine reptiles was apparent as long ago as the early nineteenth century. Mary Anning (1799–1847) was the first professional fossil collector (in Lyme Regis, England). She collected fossils from the local eroding coastal cliffs and sold them to tourists. In 1811, she came across some giant fossilized bones and sold them to Sir Everard Home, Professor of Anatomy at London University. He was unable to identify the fossil bones for what they were and it was not until seven years later that they were positively identified as coming from an extinct fish-like reptile. Although this suggestion was originally made by the famous Baron Georges Cuvier, the fossil was given its official name of *Ichthyosaurus* by W. Koenig in 1818. Many more specimens were found in the 1820s.

see also...

Cuvier

Industrial melanism

Charles Darwin's theory of evolution by natural selection lacked final proof. He could give no example of evolutionary change actually taking place yet, during his lifetime, significant changes *had* taken place due to industrial melanism. It was first demonstrated in a species of moth, the peppered moth (*Biston betularia*). Data concerning recorded changes in the populations of this species were meticulously researced by H.B.D. Kettlewell in the 1950s. These studies illustrate evolution in action, as a gradual change in form as a result of natural selection.

There are three colour forms of the peppered moth. The typical form has white wings, 'peppered' with black specks which sometimes form black lines. Another, called carbonaria, is the black or melanic form, from the black pigment called melanin. The third, an intermediate form, which is black, speckled with white, is called insularia. Before the industrial revolution, the melanic form was an extremely rare mutant. Then, as towns and their surrounding countryside became covered by soot from coal-burning factories and houses, melanic moths became more common and in some places the typical peppered variety disappeared. The soot not only blackened the trees but also killed the lichens which normally would camouflage the peppered variety, so the typical form stood out against the darker background and could be seen by predators. The black variety was now at an advantage and could survive to breed.

The phenomenon has caused an estimated 10 per cent of the 700 or so species of larger moths in Britain to have larger proportions of darker forms. In all cases the moths spend the day resting against tree bark or similar backgrounds. It must not be thought, though, that a new species has been formed in this way, because light and dark moths still interbreed to produce viable offspring. However, it has been shown that an animal's form can change as a result of selective pressures due to the environment. Where smokeless zones have been introduced since the Clean Air Act of 1956, the situation has been reversed, with the light variety once again becoming more common.

see also...

Natural selection

Insecticide resistance

In 1939, the Swiss chemist Paul Muller was the first person to use DDT as an insecticide. The compound, dichlorodiphenyltrichlorethane, was first prepared and described as long ago as 1874, but at that time its insecticidal properties were not recognized. It proved to be deadly to insects and so its use spread throughout the world. In fact, many regions that are now free of insect-borne killer diseases such as malaria and sleeping sickness owe their freedom from insect vectors to DDT. In the 1940s it was thought that in due course all harmful insects would be eliminated. However, the gene pool of insects with such large and prolific populations allowed adaptation to the selective pressures of the environment so that mutants which could cope with the insecticide were at an advantage. These survived to breed and so passed on their mutated gene to future generations. Hence populations of insecticide-resistant insects evolved: as the years passed, DDT became less effective on the housefly, for instance. Some resistance was reported as early as 1947. Today, every insect that was originally attacked by DDT has resistant varieties. With all of human technology we have not managed to eliminate even one species of unwanted insect.

The widespread use of insecticides has provided such sudden alterations in some insects' environments that fundamental adaptive changes have evolved within a few years, leading to levels of resistance that make a particular insecticide impotent. By the end of the 1960s over 225 insecticide-resistant species had been officially recognized. A large number are resistant to the cyclodiene group of chemicals (dieldrin, aldrin, lindane, etc.); next in importance is the DDT group (DDT, DDD, methoxychlor); while the organosphosphates (malathion, fenthion) are third in importance. Twenty species are resistant to other compounds which are in none of the three major groups. Many species have become resistant to insecticides of different groups. Probably the most striking example is the housefly, which, in some parts of the world, is now resistant to almost every insecticide that can safely be used.

Lamarck, Jean Chevalier de

Jean Chevalier de Lamarck (1744–1829) found fame in his lifetime, but sunk into obscurity and now his evolutionary theories are discounted. Like many sons of poor landowners, Lamarck was educated for the Church. After he was ordained, he took the name Jean Baptiste but, on his father's death when he was 17, he left college and joined the Army. On postings to Mediterranean coasts, Lamarck developed an interest in botany and he carried on with this after injury forced him to resign. *Flore Francais*, his key to plant identification published in 1778, provided evidence of Lamarck's skills, and in 1781 his appointment as a botanist to the king enabled him to travel all over Europe. Lamarck thus first made his name as a botanist but, although he worked at the Jardin du Roi from 1788, it was as a zoologist that he was to be remembered. Following the reorganization of the Jardin du Roi after the Revolution, he was given a professorship.

Lamarck specialized in invertebrates (a term he invented) and their classification, taking Linnaeus' work a stage further by dividing the 'insects' and 'worms' into groups that are still recognized in modern taxonomy. His seven-volume work on the natural history of invertebrates, published between 1815 and 1822, was the foundation of modern invertebrate biology. In fact, he also invented the name biology.

These classification studies led to observations on all animal relationships and Lamarck published ideas on this as early as 1801. The main work, *Philosophie Zoologique*, published in 1809, expressed detailed theories of animal relationships and the possibility of species changing in time – in fact, evolution. The work of Lamarck regarding evolution (Lamarckism) received little recognition in his lifetime and his successors, including Charles Darwin, regarded it with ridicule and contempt. He died a pauper and was buried in an unmarked trench. His daughter provided the poignant epitaph – 'Posterity will remember you'.

see also...

Darwinism; Lamarckism; Linnaeus

Lamarckism

In 1809, Jean Baptiste Lamarck published his most important work, *Philosophie Zoologique*. In it he tried to show that various parts of the body appeared because they were necessary, or disappeared because of disuse when variations in the environment caused a change in habit. He believed that body changes were inherited by the offspring and that, if this purpose continued for a long time, new species would eventually be produced. Hence Lamarck thought that it ought to be possible to arrange all living things in a branching series showing some species gradually changing into others. The classic Lamarckian example of a scientific error was his theory of how the giraffe developed its long neck. He maintained that the long neck evolved as each generation of giraffes stretched to reach the leaves at the tops of trees and this characteristic was passed on to future generations. Unfortunately, he never attempted to show how one species might gradually change into another on the basis of the inheritance of acquired characters. His ideas were not accepted – nor have other, more recent theories which have been proposed to explain how new species

arose. At this time also, Lamarck's contemporary, the renowned Baron Georges Cuvier, proposed his own theory of evolution. It was inferior to Lamarck's, but such was Cuvier's fame that Lamarck's ideas were ignored and remained obscure for many years.

Lamarckism fitted into the general theories of human perfectability that were prevalent during the nineteenth century. Improvement of the environment of an individual would lead, inevitably, to the improvement of the species. In the late 1890s, however, August Weismann (1834–1914) studied the nature of the reproductive cell and came to the conclusion that it was not affected by environment. He argued that the continuity of life was contained in these cells, as distinct from the body. The work of Hugo De Vries and Gregor Mendel confirmed the view that Lamarck's ideas were misleading. They served, however, to prepare the ground for the ideas of evolution soon to be conceived by Charles Darwin.

see also...

Cuvier; Darwinism; Mendelism

Lancelet, The

The first record of the discovery of the 'lancelet' was in 1774. It was found in Cornwall, England and was sent to the well-known German naturalist Peter Simon Pallas for identification. He gave it a brief description in a footnote to a book that he published and mistakenly thought that it was a sea slug, naming it *Limax lanceolatus*. Half a century later, the British naturalist Jonathan Couch found a living specimen in Cornwall and kept it in a marine aquarium where he studied its behaviour. At that time it was thought to be a primitive type of fish and he named it *Amphioxus*, which means 'sharp at both ends'. It was more commonly known as the 'lancelet'. Although now rare around the beaches of Britain, on the coast of Amoy, China, it is common enough to be used as food.

The relationship of the lancelet to the rest of the animal kingdom remains one of the most interesting features of this 10 cm long, rather drab little animal. It resembles a 'stripped-down' vertebrate in having a dorsal nerve cord lying above a supporting rod (the notochord) and an arrangement of muscles along its tail like a fish. At the same time it lacks a backbone, jaws – or indeed any bone. It also lacks anything that can be called a brain or eyes or other sense organs that can be associated with a brain. It therefore cannot be classed as a true vertebrate, yet it is as close to one as any invertebrate could be.

Its claim to being a 'living fossil' lies in the fact that there are fossils dating from the Cambrian period (540–505 million years ago) that are very similar to it. They are among many types from the Burgess Shale in the Canadian Rockies. Stephen Jay Gould focuses on the evolutionary significance of the fossil called *Pikaia* in his best-selling book *Wonderful Life*. It is suggested that *Pikaia* could be the earliest known forerunner of the vertebrates.

see also...

Burgess Shale; Living fossils

Leakey, Louis

Louis Leakey (1903–1972) was born in colonial Kenya to parents who were both missionaries. His early years were spent in Kenya but he studied anthropology at St John's College, Cambridge. Having finished his training, he devoted more than 40 years to the study of human prehistory in East Africa. His interests covered a time span from recent archaeological sites to the earliest evidence of human origins. Throughout his career, his name became virtually synonymous with the idea of an ancient origin of *Homo* in Africa.

Leakey's first fossil collecting expeditions were in the 1920s, but the beginnings of his fame among the world's anthropologists took place in 1931 at the Olduvai Gorge in Tanzania. The reason for this visit was to try to solve the mystery of 'Oldoway Man', a skeleton discovered by a German scientist (Hans Reck) in 1913. Reck believed it to be almost a million years old, but Leakey was to prove that it was in fact a 'modern' human that had been buried there. Louis Leakey and his second wife, Mary, became associated with the Olduvai Gorge and until the 1960s carried out meticulous research on ancient human use of stone tools. They found thousands of animal fossils and hundreds of stone tools during their searches.

It was Mary who found the fossil that changed their lives and professional fortunes. After working at Olduvai for 30 years, she found the fossil skull of *Zinjanthropus* in 1959. It was similar to the skull that Raymond Dart and Robert Broom had found (*Australopithecus*). Its features were generally robust, thick-boned and heavily built. The type of creature it represented was christened *Zinjanthropus boisei* or 'Boise's Man from East Africa', after Zinj, an Arabic name for East Africa, and Charles Boise, who funded the expeditions. Radiodating estimated the skull at about 1.8 million years old. The media called it 'Nutcracker man' because of its enormous jaws and teeth.

After this discovery, serious excavations began at Olduvai and continued well after Louis' death from a massive heart attack in 1972.

> ### see also...
> *Dart*

Lingula

In the absence of environmental change, it is possible to imagine a species evolving to the point where it is perfectly and completely adapted to its environment. It would remain unchanged, producing the occasional mutation – which would die out quickly because, by definition, it could only be less than perfectly suited to its environment. Such situations do actually exist. For example, today's ocean-dwelling brachiopod *Lingula* has survived unchanged for 500 million years. Some existing species are almost identical to their Cambrian and Ordovician ancestors and can be considered as 'living fossils'. The common name for *Lingula* is the lampshell, from its shell's resemblance to the shape of the oil-burning lamps of the ancient Romans.

Lampshells belong to a small group of marine invertebrates, the Brachiopods. Members of the group superficially resemble bivalve molluscs because of their two shells but that is where similarities end. They have a different feeding mechanism and life history. Living lampshells are rarely more than 5 cm long and number fewer than 300 species in almost all latitudes and at most depths in the warmer seas of the world. Most live on the continental slope but occasionally they are found on the shore. Over 30,000 species of fossil lampshells are known and some of these were nearly 30 cm long. In the Ordovician period (505–403 million years ago) they were as numerous and diverse as molluscs are today. There are whole layers of rock that are made of little else than fossil lampshells in certain parts of the world.

Apart from its simple filter feeding lifestyle, one reason for *Lingula's* remarkably long period of survival is its ability to live in conditions of low oxygen. Few animals can do this and hence it has few competitors and very few predators.

Lingula has a muscular stalk which can be shortened and lengthened and used for burrowing in mud. There are a dozen species of these in the Indo-Pacific region, especially off Japan, southern Australia and New Zealand. In some places they are used as food.

see also...
Living fossils

Linnaeus, Carolus

arolus Linnaeus (1707–1778), was born Carl von Linne in Rashult, a small town in southern Sweden. His father wanted him to enter the Church, but thought shoemaking a better career. As a boy, von Linne used to go into the country collecting specimens and was fascinated by the variety of plants and animals he found. A local doctor noticed his interest in natural history and encouraged him to go to university.

Lack of money forced him to curtail his studies in Sweden and eventually he went to Holland to complete his work. While there, he wrote the famous *Systema Naturae*, consisting of only 12 printed folio pages. This work was his 'passport' to the scientific world: scientists of repute acknowledged it, and this enabled him to travel to England and France to meet some of the most respected scientists of that time. Eventually he became Professor of Natural History at Uppsala University.

Linnaeus' great contribution to science was to name and classify plants and animals in a simple logical way using a binomial method, where organisms were given two Latin names. For example, to animals strongly resembling one another, like the lion, tiger, and leopard, he gave the overall name (Genus) *Felis*. Each would then be given a second, or trivial, name (species). So the lion is *Felis leo*, the tiger *Felis tigris* and the leopard *Felis pardus*. His love of Latin influenced von Linne to change his own name to Carolus Linnaeus in his published works.

He is often criticized as being narrow-minded and arrogant. This view was probably reinforced by his own proud motto *Deus creavit, Linnaeus disposuit* – 'God created, Linnaeus arranged'. Although he did not look beyond the external anatomy of a plant or an animal when he classified them, he directed zoological and botanical thoughts towards a logical systematic plan upon which our present science of taxonomy is based. In this way, he encouraged scientists to look at the origin of these species – a question which was to be of prime importance in the last half of the nineteenth century.

Living fossils

The Earth's fossil record shows that no species lasts for ever – the average lifespan is between one and ten million years. Living fossils are considered rare because, of all the species that have ever lived, 99.9 per cent are extinct. A living fossil can be defined as a living organism that does not differ significantly from fossils which have already been found and described. They are modern organisms with anatomical or physiological features that are normally characteristic of extinct ancestral species. They are often associated with highly restricted, remote, and almost unchanging environments and so evolve very slowly.

To this category belong the magnolia, maidenhair tree (*Ginkgo biloba*), lungfish, certain molluscs like *Nautilus*, the lampshell (*Lingula*) and the coelacanth. All 'living fossils' are not necessarily to be considered as 'missing links' but some show affinities to two major groups and hint at a common ancestry.

Examples include:

★ *Peripatus* (the 'velvet worm'),

which appears to have features in common with arthropods and annelids.
★ *Neopilina*, which may be intermediate between annelids and molluscs.
★ *Amphioxus*, the lancelet, which may link vertebrates with invertebrates.

Studies of 'living fossils' can be valuable both as a check on previous knowledge of relationships between different groups of organisms and as a means of verifying reconstructions based on fossil evidence.

Of course, extant (living today) animals and plants are not really identical to their fossil ancestors but they can be described as the remains of another day insofar as they have survived with so many features that we can identify in fossils. Sometimes they are called relict species (from the Latin *relictus*, meaning left-over).

see also...

Coelacanth; Ginkgo biloba; The lancelet; Lingula; Nautilus; Neopilina; Peripatus

63

Lizard-hipped dinosaurs

The earlier of the two dinosaur groups, the saurischians, were lizard-hipped. They had a hip structure in which the two lower bones on each side pointed in opposite directions. Some were herbivores and others were carnivores. Often they looked like bipedal crocodiles. Among meat eaters the bipedal crocodile design continued but the animals became much larger, making up the therapod group (meaning 'mammal-footed', referring to the arrangement of bones in the feet).

A typical large theropod was *Megalosaurus*, which was the first dinosaur to be scientifically described. In the Cretaceous period (144–65 million years ago), the big theropods developed in many different ways. Some had spines on their backs and, at the end of this geological period, there developed the mighty *Tyrannosaurus*, probably the best known of all the dinosaurs. Despite its 12 m length and fierce reputation, it might well have been a scavenger rather than the predator depicted in most popular films. It supported itself on powerful hind legs and had small forelimbs, all of which were equipped with huge claws. It could have been able to run fast, despite its size.

There was a much smaller type of theropod that hunted small prey, some of which were the size of a chicken. Close relatives of these might well have been warm blooded – particularly as they reputedly gave rise to birds. In the Jurassic and Cretaceous periods, another lizard-hipped dinosaur subgroup appeared, members of which were called sauropods. These became the true 'heavyweights' of the dinosaur group and were all quadrupedal and herbivorous. They had lizard-like arrangements of bones in their feet and were truly massive, with tiny heads, long necks and long tails. *Apatosaurus*, sometimes called *Brontosaurus*, was one of the largest, weighing in at 30 tonnes and a height of 12 m. It lived in swamps or on the shores of large bodies of water, probably using the water to support its great weight. It is possible that the huge volume of the sauropods, together with their slow metabolism, enabled them to develop a constant body temperature.

see also...

Bird-hipped dinosaurs; Dinosaurs

'Lucy'

On Christmas Eve 1974, Donald Johanson and his colleague Tom Gray found a piece of hominid arm bone jutting from a rocky slope at Hadar in Ethiopia. Other pieces of bone were discovered nearby, and soon they realized that they had an amazing find – a partial skeleton. After another three weeks of searching, about 40 per cent of a female hominid skeleton was recovered. Officially catalogued as AL 288-1 Partial Skeleton, it was named *Australopithecus afarensis* – 'southern ape from afar'. However, she is better known as 'Lucy' because her discoverers were so elated at finding her remains that they celebrated by drinking beer and listening to an old Beatles favourite, *Lucy in the sky with Diamonds*, well into the night – she was named after the song.

'Lucy' had lived just over 3 million years ago. She is the most complete specimen of *Australopithecus afarensis* ever to have been found. Detailed reconstructions from the remains indicated that she was about 120 cm tall –the equivalent of a six-year-old child of today. She weighed around 60 pounds, which is less than an adult chimpanzee. From her teeth it was estimated that she had died in her late teens or early twenties because one of her molars was present in her jaw bone. Her gender was determined by the properties of her hip girdle. From the waist down, 'Lucy' appears to be human-like with legs and hips suggesting the possibility of bipedal walking. Above the waist, she appears more ape-like, with long arms and a sturdy rib cage, hinting at tree-climbing behaviour. Most palaeontologists classify hominids within one group called the Southern apes (*Australopithecus*). They resemble chimpanzees but human features include the spinal cord extending from below the brain rather than pointing backwards like those of apes. The teeth show a mixture of ape-like and human-like forms.

'Lucy' was believed to be the original and earliest hominid until 1994 when an even older species, *Ardipithecus ramidis*, was found. This could be 4.4 million years old.

Lyell, Charles

Charles Lyell (1797–1875) began his career as a barrister, studying at Oxford and then at Lincoln's Inn before being called to the bar in 1825. However, while he was at Oxford, his interest in geology had been stimulated by William Buckland's lectures and he practised as a barrister for only two years. Even before this, in 1819, he was elected a fellow of the Linnaean and Geological socieites and, apart from his short time at the bar, devoted the whole of his life to geology.

After working on tertiary deposits in France and Italy, Lyell wrote his major work, *Principles of Geology*. In this he popularized the Principle of Uniformitarianism. This work profoundly influenced Charles Darwin in his formulation of ideas on natural selection and evolution. In fact, Lyell was one of the people who encouraged Darwin to publish the *Origin of Species* in 1859.

In 1835 Lyell was made President of the Geology Society. Three years later he finished *The Elements of Geology*, which soon became a standard work on stratigraphy and palaeontology. He was an opponent of Cuvier's hypothesis of catastrophism, his argument being that the forces shaping the surface of the Earth (wind, glacier, erosion, earthquakes and volcanoes) operate in the present as in the past and that these forces continue to produce their effects over long periods of time.

Lyell travelled widely throughout Europe and America and wrote several books based on his work in these countries. In 1848 he was knighted and received a baronetcy in 1864. A man of theory, Lyell (through his writings) achieved world-wide influence in his field, and the Principle of Uniformitarianism, which he adopted and publicised, became a cornerstone of geological thought. He found many inconsistencies in the ideas of Lamarck, being especially at odds with Lamarck's assumption that species were continuously changing.

see also...

Cuvier; Darwin; Darwinism; Lamarckism; Natural selection

Mantell, Gideon

One of the first people to find proof that prehistoric reptiles had existed was Gideon Mantell (1790–1852). He was a physician and a keen amateur palaeontologist. Legend has it that one Spring day in 1822, Dr Mantell visited a patient with his wife, Mary Ann. While he was seeing the patient, his wife went for a stroll along a road that was being repaired. She noticed something shining in a pile of stone that was being used for road repair. On closer examination, the shiny objects turned out to be fossil teeth. Gideon Mantell had not seen anything like these before. He deduced that they were from a herbivore because they were blunt and had been worn down by constant chewing. The only things he could compare them with were teeth of elephants or rhinoceroses. With the help of the stone suppliers, he traced the source of the fossils to a quarry in the Tilgate Forest. However, the age of the particular rocks in this quarry was about 130 million years old, and this pre-dated large mammals by many millions of years. The teeth came from the Mesozoic era – the age of the reptiles.

Mantell sent the teeth and some fossil foot bones that he later found in the quarry to two of the most famous authorities of the day: Baron Georges Cuvier in Paris and Dr William Buckland, the Professor of Geology at Oxford University. These two scientists' responses were not encouraging, as neither thought that they were from extinct reptiles. Despite this reaction, Mantell persisted with his hypothesis and searched through many collections of ancient and modern skeletal remains in museums for anything that looked remotely like his specimens. At last, after three years, he was shown a skeleton of a modern South American iguana at the Hunterian Museum of the Royal College of Surgeons in London. In the lower jaw he recognized mini-versions of the teeth that his wife had found quite by accident. He published a description of the find and called it *Iguanodon*.

see also...

Cuvier

Mendelism

Classical genetics are named after Gregor Mendel. Mendelism is the study of inheritance by controlled breeding experiments, first carried out by Mendel in the 1860s. The characteristics studied are usually controlled by one gene and show a simple dominant or recessive relationship between alleles. Large numbers of progeny from a given cross are counted to find the ratios of various phenotypes (the outward expression of the genes, which may be the outward appearance) and from this the parental genotypes can be assessed. Work of this nature gave the first indication that inheritance is particulate rather than blending.

Mendel formulated two laws to explain the pattern of inheritance he observed in crosses involving the common garden pea, *Pisum sativum*. The first law, the Law of Segregation, states that any character exists as two factors, both of which are found in the body cells (somatic cells) but only one of which is passed on to any one gamete. The second law, the Law of Independent Assortment, states that the distribution of such factors to the gametes is random. Therefore, if a number of pairs of factors is considered, each pair segregates independently.

Mendel gave the name 'germinal units' to those features that he considered to control characteristics in his experimental pea plants. Today these germinal factors are called genes (a term first used by the Dane Wilhelm Johannsen in 1909). The different forms of genes are called alleles, an abbreviation of 'allelomorphic pairs of genes'. It is known that a cell with the full complement of chromosomes for a species (diploid cell) contains two alleles of any particular gene. Each allele is located on one of a pair of homologous chromosomes (chromosomes that pair during sex cell formation). Only one homologue of each pair is passed on to a gamete (an egg or a sperm). Thus, the Law of Segregation holds true. Mendel envisaged his factors as discrete particles but it is now known that they are linked together on chromosomes. The Law of Independent Assortment therefore applies only to pairs of alleles found on different chromosomes.

Mutagens

Mutagens are changes in the external environment which can have profound effects on mutation rates.

Radiation

Organisms are constantly subjected to various types of radiation. The electromagnetic spectrum extends from long radio waves to extremely short cosmic rays. The amount of energy contained in the radiation becomes larger as the wavelength becomes shorter, and above a certain energy level the rays can penetrate living cells.

Ultraviolet radiation can penetrate cells less well than higher energy rays but is readily absorbed by DNA, causing structural damage at a molecular level and skin cancer can result. Radiation of wavelengths shorter than that found in ultraviolet rays is called ionizing radiation. The energy level is so high that electrons in the irradiated atoms may be knocked out of orbit, thus producing a positively charged ion. Ions, and the molecules which contain them, are chemically much more reactive than the original neutral atoms. The structure of DNA, and hence

chromosomes, can be affected by this type of radiation but any mutations caused will not be maintained in the population unless they occur in organs that produce sex cells.

All organisms are subjected to low levels of ionizing radiation from cosmic rays and from radioactive materials found in the Earth's surface rocks. We are also likely to encounter additional radiation from human uses of radioactive isotopes, such as medical X-rays and radioactive fall-out from nuclear accidents.

Chemical

Since 1945, a long list of mutagenic chemicals has been compiled and these often have carcinogenic (cancer-causing) properties. The use of mutagens has proved valuable in artificially selecting new variants in horticulture and agriculture. One of the most important of these is colchicine, which causes the doubling of chromosomes in cells and produces larger than normal plants.

see also...

*Artificial selection in plants;
Mutation*

Mutation

Mutations are abrupt, hereditable changes in single genes or regions of a chromosome. Although the term mutation is now generally used in this sense, historically it was defined more broadly to include alterations in chromosome number and chromosome structure. Mutations constitute the raw material of evolution and are the basis for the variability in a population on which natural (or artificial) selection acts to preserve those combinations of genes best adapted to the environment.

Many mutations may be neutral or 'silent' (i.e. they have no observable effect on the organism). Harmful mutations become evident because they may alter the survival capacity of the organism. Mutations occur randomly and spontaneously and may also be induced by environmental factors. Spontaneous mutations arise from errors in replication and different genes may mutate at different rates.

Induced mutations can be due to exposure to environmental factors (mutagens) such as

★ **X-rays**. These cause breaks in the DNA leading to chromosomal rearrangements (block mutations) and deletions.
★ **Ultra violet rays** cause point mutations including base-pair substitutions, insertions and deletions.
★ **Chemical agents**. These include:
— Base analogues, which mimic and substitute for normal bases in DNA synthesis, leading to mispairing.
— Reactive chemicals, which add chemical groups to or delete them from normal bases, leading to mispairing during DNA replication (examples are benzene, formalin and carbon tetrachloride).

Genes mutate at known rates, but the rate varies depending on the gene involved – some have high spontaneous mutation rates. Most mutations occur in somatic (body) cells and are not inherited. Those that occur in sex cells may be inherited.

see also...

Mutagens

Natural selection

The process which Charles Darwin called the 'struggle for survival' by which organisms less adapted to their environment perish, and better adapted organisms tend to survive. According to Darwinism, natural selection acting on a varied population results in evolution. Natural selection, then, came to be defined as the process through which certain types of organisms are more reproductively successful than other types, thereby disproportionately passing along those traits that led to their success. Darwin envisaged nature determining the reproductive success of individuals. The traits of those individuals with some reproductive advantage could be expected to increase through generations.

The sheer logic of Darwin's theory put forward a unifying concept which accounted for his observations. In fact, he used three observations and two deductions based on them to formulate his theory:

★ The first observation was stated in Malthus's *Essays on Population*: organisms have a tendency to multiply in geometric progression – 2, 4, 8, 16, 32 and so on.

★ The second observation was that the numbers of a species tend to remain constant over long periods of time.

★ The first deduction was based on these two observations: a constant struggle for existence takes place. Organisms are in constant competition for the chance to reproduce.

★ The third observation was that all living things vary.

★ The second deduction was natural selection: some individuals are more likely to succeed in reproducing than others. Those with favourable characteristics will be more likely to survive and reproduce than those with unfavourable characteristics.

Darwin's theory was not to be rejected on any basis other than a better explanation.

> **see also...**
>
> *Darwinism; Neo-Darwinism*

Natural selection by predation

Charles Darwin was aware of the importance of population size in determining the survival of individuals competing for limited essentials such as food. In this process, predation must play an important part. There have been many studies of predators acting as selective agents, most notably those of the land snails of the genus *Cepaea*. Only two species of *Cepaea* occur in Britain, the brown-lipped snail, *C. nemoralis*, and the white-lipped snail, *C. hortensis*. The brown-lipped type is the more variable of the two, the colour of its shell being classifiable into three distinct shades – brown, pink and yellow. The markings on the shell also vary, from unmarked forms to those with five dark bands, with a number of intermediates. The white-lipped type is usually yellow and there are either no markings or five distinct bands: intermediates are relatively rare. In both species, shell colour and pattern have been shown to be under genetic control.

Studies of the brown-lipped types in a variety of habitats such as grassland and woods showed that there was a high correlation between yellow shells and green backgrounds such as downland. In darker places, ranging from shaded hedgerows to deepest beechwoods, the proportion of yellow gradually declined almost to zero. Again, banding on the shells showed a clear correlation with the environment, the more varied and dark the environment (e.g. woodland) the greater the proportion of banding.

The main predator of the brown-lipped snail is the song thrush. In mid-April, the vegetation is mainly brown and yellow shells are at a selective disadvantage. By late April, yellow is of neutral survival value relative to the other colours, but by mid-May it is at an advantage.

There is clear visual selection by predators – certain shell patterns seem to provide camouflage against certain backgrounds. However, the white-lipped type also has variable shell patterning and may live in the same locality as the brown-lipped snail. They are hunted by the same predators but their gene frequencies may be very different. In beech woods almost all the brown-lipped are unbanded browns whereas the white-lipped are mostly banded yellows.

Nautilus

The chief interest of *Nautilus* as a living fossil lies in the fact that it is the only living representative of a vast multitude of cephalopods with external chambered shells (ammonites) which flourished between the earliest Cambrian period and the late Cretaceous period (475 million years ago). This expanse of time embraced the longest part of the history of the life of invertebrates on Earth. After being the dominant type of marine invertebrates in the Mesozoic era (245–65 million years ago), they suddenly became extinct, and the cephalopods are now mainly represented by the octopus, squid, and cuttlefish.

The ammonites are named after the ancient Greek god Zeus Ammon, because of their resemblance to the god's spiral horns. Thousands of extinct species are known by their well-preserved fossils. From the beginning of the Triassic period (245 million years ago) onward, new families, genera and species evolved rapidly. These are differentiated by the shape and sculpture of the shell whorls, but particularly by the patterns at the junctions of the shell whorls.

Nautilus exists today as at least six species and lives at moderate depths on some tropical coasts in the southwest Pacific. It swims, mainly at night, by a form of jet propulsion. Although the shell is so heavy relative to the body, it is buoyed up by its gas-filled chambers. Because the centre of buoyancy is lower than its centre of gravity, the nautilus is always body downmost, just as the basket of a balloon is always lower than the balloon itself.

Although the shell of the nautilus has been known since the sixteenth century, the animal itself was not properly studied until 1831 when Richard Owen of the Royal College of Surgeons, London, described a preserved one which had been brought to him from the Pacific.

Over 100 years lapsed before living specimens were studied. In 1962, Dr Anna Bidder of Cambridge University visited New Caledonia, where she studied *Nautilus* in aquaria. Since then, they have been filmed in their natural environment with the use of submersible craft.

see also...

Living fossils; Owen

Neo-Darwinism

Neo-Darwinism is Charles Darwin's theory of evolution through natural selection, modified and expanded by modern genetic studies arising from the work of Gregor Mendel. Such studies have answered many of the questions that Darwin's theory raised but could not adequately explain because of lack of knowledge of genetics at the time it was formulated. The essence of neo-Darwinism is that the adaptation of organisms to their environment is brought about by natural selection acting upon small inherited variations, most of which are initially non-adaptive. Mutation is the ultimate source of a new variation and a mutation is preserved and transmitted by the mechanism of particulate inheritance. Changes to the gene frequencies of different populations leading to the formation of varieties and, eventually, new species depend upon various forms of reproductive isolation and the restriction of gene flow. Although selection pressures are often severe, the formation of new species is essentially a slow process.

The critics of neo-Darwinism maintain that it fails to account for discontinuity and the outbursts of rapid evolutionary activity that have alternated with those of inactivity. The latter could be explained by periods of ecological stability alternating with those of rapid environmental change. However, the inherent slowness of a neo-Darwinian mechanism has prompted the idea that large mutations involving much larger phenotypic changes may also have played a part.

Superimposed on the idea of sudden major changes in the evolutionary pattern is the problem of intermediate stages. Such stages must have preceded many adaptations that eventually proved successful. For example, the explanation of a 'half-evolved eye' presupposes that it was preceded by a range of devices that concentrated light in varying degree but did not form an image. However, any structure that increased the capacity to detect light at varying intensity would have been an advantage in the perception of shadow and movement and would aid survival.

see also...

Darwinism; Mendelism; Mutation; Reproductive isolation

Neopilina

The molluscs (snails, clams, squids and slugs) are highly successful and widespread today, probably because their ancestors evolved the effective protection of a shell. One of the most primitive molluscs which may resemble the ancestor of the whole group is *Pilina*, appearing in the Cambrian period with a simple cup-shaped shell. Such forms were thought to have become extinct about 350 million years ago but in 1952 living specimens of a species (now called *Neopilina*) were dredged from the depths of the Pacific. To the non-specialist, *Neopilina* is a somewhat boring-looking animal – just another limpet about 2 cm in diameter – but to the experts in the world of malacology its discovery was one of the most exciting molluscan events of the twentieth century. It has gills in pairs and a bilaterally symmetrical body (distinct right and left halves), which is unusual for molluscs but a common feature in annelids. Apart from the fact that it is a 'living fossil', showing possible links between the annelids (true ringed worms) and the molluscs, the whole circumstances of its discovery are astonishing.

In 1952, the Danish deep-sea research ship *Galathea* dredged up specimens from 3600 m off the coast of Costa Rica. Nothing quite like them had been seen alive before, and because they had been brought up from such depths it was not possible to do more than speculate on their way of living. Six years later a second species was found off Peru; another discovered off California in 1962; and then, only nine years later, a fourth species was collected in the Gulf of Aden. From these data it is fair to assume that these animals have a very wide distribution in the deep oceans of the world. The amazing point about finding *Neopilina* is that it had remained undiscovered for so long. Since 1872, with the three-year epic voyage of HMS *Challenger* that marked the birth of oceanography, there have been scores of voyages by deep-water research vessels, some covering vast areas with intensive dredging: 80 years of searching for new marine species had missed *Neopilina* completely.

see also...

Living fossils

Neotony

The form eventually attained by the adults of a species is controlled genetically and can be subject to natural selection during the process of growth. The genes involved determine both the nature of the adult characters and the rate at which they appear, including the onset of sexual maturity. The speeding up of the development of sexual maturity relative to body form can result in an adult organism retaining juvenile characteristics. This is called neotony – the precocious attainment of sexual maturity in larval stages. A classic example of the influence of neotony on evolution was suggested by Walter Garstang and others in the 1920s. They showed that the chordates (animals with a notochord, and the ancestors of vertebrates) are probably derived from the sexually mature larvae of echinoderms (spiny-skinned invertebrates) like the sea urchins.

There are many recorded instances where the onset of neotony has played an important part in evolution, particularly in primitive amphibians. One of the best known examples is the Mexican axolotl. It has been described as 'the Peter Pan of the amphibians', being able to reproduce while still in its aquatic larval form; it can complete its life cycle without ever leaving the water. Other amphibian examples are the olms of Europe and the mud puppies of the USA.

These obscure examples, although interesting zoologically, are probably outside the experience of most people. However, a fascinating idea was put forward by B.G. Campbell in the 1970s, which certainly *is* within the experience of all of us. He applied the phenomenon of neotony to humans. Our limited body hair and the fair skin of Caucasoid races may well be the result of the prematurely arrested development of the hair follicles and the melanin pigment cells associated with them. Again, the relatively large brain and absence of brow ridges are characteristic of the *young* stages of apes; as is the rate of growth of the brain compared with that of the rest of the body and the enormous size of the cerebral hemispheres at birth in comparison with that of neighbouring organs.

see also...

Natural selection

Niche

The word is derived from the Italian *nicchia*, which means a cavity into which something fits. In ecology, the meaning has been extended to mean the 'occupation' of an organism. It is the total functional role of an organism in a community. The niche encompasses all the bonds between the population, the community, and ecosystem in which an organism is found. These bonds include the following adaptations:

★ tolerance ranges for all abiotic environmental limiting factors
★ the ability to exploit types of food within the environment
★ modifications allowing survival in areas in which the organism lives
★ the ability to survive within a particular population structure

Every population has an ecological niche, which is the major determinant of the structural, physical and behavioural adaptions of that population. Different communities in ecosystems characterized by similar environments are often very similar in their structure and may contain one or more niches that are essentially identical. The adaptations of populations inhabiting these niches may also be very similar in appearance, even though the populations are totally unrelated.

If two species occupy the same niche then competition occurs until one species has replaced the other. A similar niche may be occupied by different species in different areas. For example, the fallow deer of Africa occupes the same niche as the red deer of Eurasia. Conversely, one type of organism may evolve by adaptive radiation to fill several niches, such as Darwin's finches of the Galapagos Islands.

Where two species succeed in co-existing, we can assume that they are interacting with the environment in different ways (occupying different niches). This can be illustrated while walking through a wood and seeing various species of seed-eating birds in the same area: they are, in fact, using the habitat differently.

see also...

Darwin's finches; The Galapagos Islands

Origin of life

The main theories that have been proposed can be considered as variants of four main hypotheses:

1   Life has no origin. Life, matter and energy are co-existent in an infinite and eternal universe. According to this theory, 'seeds' of life travel from one planet to another through space.
2   Life was created by a supernatural event at a particular instant of time in the past.
3   Life arrived on this planet from elsewhere in the solar system or universe.
4   Life arose on this planet strictly according to chemical and physical laws as we know them.

Of these theories, 1 is not consistent with the facts as they are known today, and 2 is not capable of scientific investigation as it rests completely on faith. Theory 3 is unsatisfactory in that it merely refers the problem to some other point in space and time, and is unlikely because of the almost insuperable difficulties involved in crossing interstellar or interplanetary space. These difficulties include the necessity for complete suspension of all metabolic processes and the more serious effects of cosmic radiation met in space. These factors would prove lethal to any living organism which was unprotected. This last criticism also applies to the 'seeds' invoked in the first theory.

The fourth theory is the most plausible one to investigate. In 1936, the Russian biochemist A.I. Oparin published *The Origin of Life*, in which he proposed that the atmosphere of early Earth was a mixture of ammonia, water and methane. In 1953, Stanley Lloyd Miller simulated this atmosphere in an apparatus through which he passed a continuous electric spark for one week. In this way he made types of amino acids (the building blocks of proteins) from inorganic molecules. By 1968, every known amino acid had been synthesized in a similar fashion. By the early 1990s all the basic building blocks of DNA had been synthesized from simple chemicals and, as the end of the twentieth century approached, self-replicating molecules were synthesized.

Owen, Richard

Richard Owen (1804–92) was a comparative anatomist and was born in Lancaster. He studied medicine at Edinburgh University and then at St Bartholomew's Hospital. On advice from his superiors, Owen concentrated on scientific research and worked in the museum of the Royal College of Surgeons, compiling many catalogues. In 1832, he published a treatise on the Pearly Nautilus and devoted the next 50 years to studying and writing about the comparative anatomy of many animals, particularly birds. He was able to apply his anatomical knowledge to the difficult problem of reconstructing extinct forms from fragmentary fossil material and made important contributions to palaeontology and the study of evolution. He became an expert anatomist, gaining his skill by dissecting animals that had died at London Zoo. His reconstructions of extinct forms of reptiles from fossil bones often resembled gigantic versions of known modern types.

Owen's main claim to fame as a contributor to the study of evolution is his invention of the name 'dinosaur'. The name was first coined by him in 1842 from two Greek words, *deinos* ('terrible') and *sauros* ('lizard'). On New Year's Eve, 1853, he and 21 other scientists of repute dined inside one of Owen's massive models of a dinosaur in the grounds of the Crystal Palace, London.

In 1849 he became conservator of the museum of the Royal College of Surgeons, and in 1856 superintendent of the Natural History Department of the British Museum. In this capacity he brought about the setting up of a separate Natural History Museum in South Kensington, London, and spent twice the annual acquisition budget on the first almost complete fossil of *Archaeopteryx* ever to be found. Owen remained in this office until 1884 when he retired to Richmond.

Although he was most industrious, some of his work has proved incorrect. A prolific writer on comparative anatomy, Owen was also the pioneer of vertebrate palaeontology. Opposed to Charles Darwin's theory of evolution, he resisted the idea of humans having a common ancestor with higher primates.

see also...

Archaeopteryx; Darwinism; Dinosaurs

Peking Man

Peking Man is a recent branch of human ancestry, with fossil remains dated about 1.8 million years old. The first evidence of his existence came in the form of an ancient fossil tooth, which was purchased in a Peking apothecary in 1903 by the celebrated German palaeontologist Max Schlosser. During the Geological Survey of China in 1921, palaeontologists from at least seven different countries hunted for the fossils of early humans near the village of Choukoutien (now Zhoukoudian). Among those dedicated anthropologists were the Swedish J. Gunnar Andersson and the Canadian Davidson Black. In 1926, they found two fossil teeth in an abandoned quarry near Peking and identified them as having come from Peking Man. This early ancestral branch of the human line of evolution was given its scientific name by Black: *Sinanthropus pekinensis*, which means 'Chinese man of Peking'. Between 1927 and 1937, palaeontologists from the Chinese–Swedish survey found remains of more than 40 individuals, with evidence of cannibalism and the use of fire and primitive quartz tools. Black died in 1934, and his successor at the University of Peking (the German anatomist Franz Weidenreich) found many more fossils of Peking Man between 1936 and 1937.

With the approach of the Second World War, the transport of the fossils of Peking Man to the USA was arranged. Unfortunately, on the very day that the fossils were due to be shipped to America, the Japanese bombed Pearl Harbor and war was declared between the USA and Japan. Japanese troops intercepted the train on which the fossils were being transported to the port of Chinwangtao (now Quinhuangdao) and they were never seen again. It was fortunate that Weidenreich had made plaster casts of the fossils and these were transported to the USA. A long search for the original priceless fossils had ensued since the end of the war but without success. Fifteen years after Peking Man's naming as *Sinanthropus pekinensis* the name was changed to *Homo erectus*, the same as Java man found by Eugene Dubois.

see also...

Dubois

Pentadactyl limb

Tracing the patterns of homologous structures found in different organisms can show how the structures have become adapted to fulfil different functions. A classic example is the five-fingered (or pentadactyl) limb of vertebrates. It evolved from modifications to the paired paddle-like fins of ancestral lobe-fin fishes (Crossopterygii) and the basic pattern of the bones can be traced through amphibians, reptiles, birds and mammals. Various alterations have been made to this basic limb type by the reduction or fusion of elements as an adaptation for different functions and modes of progression. The limbs may be specialized for swimming, as in the flippers of the whale or the wings of penguins; for walking, as in the legs of terrestrial vertebrates; for manipulation, as in the fore limbs of primates (monkeys, apes, and humans); for flying, as in the wings of birds and bats. The bones originated from the same ancestral stock in all of these cases, and are shown in the table opposite.

Adaptation for the same mode of life may be achieved by different means; bird wings, for example, depend on

Fore limb	Hind limb
Humerus (upper arm)	Femur (thigh bone)
Radius and ulna (lower arm)	Tibia and fibula (shin bone)
Carpals (wrist)	Tarsals (ankle)
Metacarpals (hand)	Metatarsals (foot)
Phalanges (fingers)	Phalanges (toes)

feathers whereas bat wings use a membrane of skin. The skeleton beneath, however, is remarkably similar. This similarity suggests that they have both evolved from some common ancestral terrestrial vertebrate. The appearance of flight in mammals was a very much later development than flight in birds. Bats evolved to exploit a niche unfilled by other mammals and left unoccupied after the demise of the flying reptiles, the pterosaurs. Flight in birds and bats does not demonstrate recent common ancestry between the two – the wings have evolved independently in each but from the same basic starting point in both cases: the pentadactyl limb.

see also...

Niche; Pterosaurs

Peripatus

In rocks of the Cambrian period (540–505 million years old), fossils of animals have been found which seem to link the Annelida (true ringed worms, such as the earthworm) with a group of jointed-legged invertebrates, the Arthropoda. Such a fossil is *Aysheaia* of the Burgess Shale. This small marine animal was ringed like the worms but it had walking legs, more typical of arthropods like centipedes of today. Their descendents invaded land and possibly gave rise to the present-day relict species ('living fossil') called *Peripatus*.

Peripatus closely resembles its Cambrian ancestor and survives today as 70 species throughout parts of tropical America, Africa and Australasia. Its body is worm-like, tapering towards the hind end. It is about 7 cm long but can extend or contract. The colour can range from dark slate to reddish brown. The skin is dry and velvety to the touch and it has 20 or so pairs of legs, each ending in a pair of hooks and ringed like its body. The most remarkable thing of all is that it has become totally land-living but still has a breathing system which ties it to damp areas.

As long ago as 1874, it was decided that *Peripatus* was more closely related to arthropods than to annelids, when H.N. Mosely became the first qualified naturalist to examine living specimens. Ironically, they were found in a forest habitat even though Mosely was on an oceanographic survey at the time, on HMS *Challenger*. He took advantage of the ship's seven-week refit at Cape Town and explored the local area – hence the first officially recorded species was named *Peripatus moseleyi*, after Mosely.

The fossil ancestor, *Aysheaia*, is over 500 million years old. Although 19 specimens of this fossil have been found, the first was little more than a rusty-coloured stain in a piece of limestone. Yet it was enough to convince the specialists that the structure of its body and legs relate it, albeit distantly, to *Peripatus*, commonly known as the 'velvet worm', which is still crawling around in some damp tropical forests.

see also...

Burgess Shale; Living fossils

Phenetics

henetics is a name given to a method of classification of organisms which is quite different from cladistics. It relies on the observable similarities and differences between organisms that can be used to indicate natural, rather than evolutionary, relationships between groups. Prime supporters of the idea are Peter Sneath and Robert Sokal, who published their research on this topic in the 1970s. They state that it is impossible to classify organisms objectively according to genealogy (tracing ancestral patterns). Instead they investigate similarities and differences in the outward appearance of organisms and compare them. To construct phenetic classifications, sophisticated mathematical tools and models are needed and so this type of classification (now often called numerical taxonomy) has become more important as the use of computers has developed. A large number of characters are measured in the different organisms being studied and coefficients of overall similarity are calculated. The organisms are then placed in pairs with the highest coefficients of similarity; the pairs formed are then compared with each

other and so on until all the pairs of characteristics have been considered.

One of the main claims of the pheneticists is that this method of classification eliminates the need for making subjective judgements. The characteristics chosen are all measurable and so can be dealt with objectively. This raises the problem of *which* characteristics should be chosen to measure: any judgement here may be subjective.

One flaw in this method is that, depending as it does on external features, it may also group together organisms that show convergent evolution rather than close ancestral relationships.

It is possible that, with the introduction of genetic profiling to classification, any method of classification based solely on external features will become redundant very quickly.

see also...

Cladistics; Convergent evolution; Genetic profiling

Piltdown Man

The Piltdown Man was one of the most notorious hoaxes ever to have happened in anthropology. It dates back to 1912 and was not recognized as a hoax until 1953. The skull of Piltdown Man was reconstructed from fossils found in a gravel bed at Piltdown Common in Sussex, England. The remains were discovered by Charles Dawson and the Keeper of Geology at London's Natural History Museum, Arthur Smith Woodward. They found nine pieces of the brain case of the skull, the right part of a lower jaw with two molar teeth, some flint tools and the fossil bones of some extinct animals after a thorough search of the site. The remains of the skull were thought to be 200,000 years old.

The media called Piltdown Man the 'missing link' and it was officially given the scientific name of *Eoanthropus dawsoni* – 'Dawn Man of Dawson'. During the next 30 years ancient hominid fossils were found in Africa and Asia but none were very similar to the Piltdown specimen. In 1948 the Piltdown remains were subjected to the newly developed technique of fluorine dating and serious questions were raised about the authenticity of 'Piltdown Man'. It turned out that the jaw was from a medieval orang-utan that had died at about the age of 10. The skull pieces were older than the jaw but not of significant age. Some of the teeth had been filed down to look worn. So it was proved beyond doubt to have been a well-planned hoax.

Piltdown Man fitted well with what people had hoped for at the time – an evolutionary 'missing link' combining features of humans with those of apes. It showed that sometimes preconceived ideas and hopes can override logic and good observation. The perpetrator of the mischievous deed was never found (and probably never will be). Joseph Sidney Weiner applied some forensic investigations to the episode in 1954 and wrote up his findings in his book *The Piltdown Forgery*. Despite Weiner's meticulous research, the Piltdown fraud still remains a mystery.

see also...

'Lucy'; Peking Man

Plesiosaurs

large group of swimming reptiles from the Triassic and Cretaceous periods (245–144 million years ago) was the plesiosaurs. Two main lines of these animals developed from a common ancestor in Triassic times, both very different from each other. Their ancestors were something like the 3 m long *Nothosaurus*, shore-dwelling fish eaters with long toothy jaws, long necks, webbed feet and a fin along the top of their long tail.

The plesiosaurs were far more specialized for sea life but not as advanced as the ichthyosaurs. The elasmosaurs made up one major group and had long necks with squat bodies. Their limbs were developed into paddles that drove them through the water with a 'flying' action. Steering was helped by a diamond-shaped paddle on the end of a short tail. They probably fished the waters of the ocean surface, using their long necks to catch prey by jabbing rapidly among the scattering schools of fish. In this way they could have seized several fish at a time without moving the body at all.

An early discovery of a fossilized skeleton of *Elasmosaurus* in the USA in 1868 began a remarkable feud between two rival palaeontologists, Edward Cope and Othniel Marsh. The specimen was described by Cope, who mistakenly put the skull at the end of the tail instead of at the end of its neck. Marsh pointed this out to him and to the general public, and it is said that Cope never forgave Marsh for his public humiliation.

The short-necked line of aquatic reptiles, the pliosaurs, were much larger than the elasmosaurs. Some, such as *Kronosaurus* from the Cretaceous of Australia, were about 12 m long. They had very large heads, up to one-fifth of their body length, and were streamlined. They must have represented the reptilian equivalent of the sperm whale of today, fishing for the ancestors of the modern giant squid in the depths of the ocean. Ichthyosaurs, elasmosaurs and pliosaurs probably did not compete with each other for food because they could have fished at different depths in the ancient seas.

see also...

Ichthyosaurs

Primate origins

Primates evolved from an offshoot of the insectivores (insect-eating mammals like shrews). Most of today's primates display similar characteristics, including a highly developed brain, good eyesight, a relatively poor sense of smell and hands useful for grasping. Only among New World monkeys (from South America) are prehensile tails found: in the larger Old World apes (from Africa and Asia), such as the gorilla, the tail has been lost.

The 'stock' ancestral primate is thought to be *Plesiadapis*, which lived in North America and Europe about 60 million years ago. It had rodent-like features, such as gnawing teeth, and probably looked very much like a modern squirrel. The face had a long snout and the sense of smell was likely to have been very important, while the brain was small and simple in structure. The skeleton was also primitive, with front and back legs of about equal length and claws on all toes, which had no grasping capabilities.

About 50 million years ago, a group of more advanced primates developed in North America and Europe. Members of the group resembled modern lemurs in many more ways than did the earlier *Plesiadapis* – for example, in the shape of the teeth, relatively long hind legs, larger brains, reduced snouts, and eyes that faced forwards. They differed in other ways, having a greater number of teeth and lacking the grooming comb formed by the front teeth, which is found in many modern lower primates.

They differed, also, in the way they moved through trees. Many modern lower primates move around by 'vertical clinging and leaping', clinging to upright tree trunks and springing off using their powerful back legs. Their ancestors did not evolve this specialization: they were four-footed climbers which moved through the smaller branches of forest trees. However, they had to widen their diet to include fruits and buds as well as insects.

None of these fossil types are likely to be the ancestors of the higher primates (monkeys and apes). Another widespread Eocene group, the Omomyids, were relatively unspecialized and were probably the ancestors of the higher primates.

Proconsul

In 1927, H.L. Gordon, a settler in western Kenya, found some unusual fossils embedded in limestone while he was digging in a quarry. He sent them to the British Museum for identification, and when they were separated from the surrounding rock they were recognized as part of the upper jaw bone and tooth of an ancient hominoid. Other fossils from the same site had been dated at around 18 million years old, (fairly early in the Miocene period). The identification and description were made by Tindell Hopwood, a palaeontologist at the British Museum. He became most interested in these particular specimens because ape fossils were very rare and Gordon's were older than any he had seen. In 1931, he organized an expedition to the site and found two more fossils. Hopwood became convinced that he had found the remains of the ancestor of modern chimpanzees, which gained the scientific name of *Proconsul africanus*. With some sense of whimsical humour, Hopwood had named *Proconsul* after Consul, a chimpanzee in a London vaudeville act that entertained audiences by riding a bicycle on stage while smoking a pipe.

More fossils of *Proconsul* were found in 1948 by Louis and Mary Leakey; parts of the face and jaws on Rsinga Island in Lake Victoria. Parts of a skull, arm, foot, and hand were found in 1951 at the same site which added to a more complete skeleton. By the 1990s, parts of nine *Proconsul* skeletons had been found: adults, juveniles, and babies. Almost every bone in *Proconsul*'s body is now known.

The current evidence indicates that there were possibly three species of *Proconsul* living about 18 million years ago, and that it was a remarkably generalized ape. It did not seem to have any adaptations for leaping or for swinging by its arms, although it could have 'knuckle-walked' as do chimpanzees and gorillas.

Analysis of *Proconsul*'s abundant remains have indicated that it was *not* an ancestor of modern chimpanzees as Hopwood thought. Instead, it might be a generalized ancestor of all the larger anthropoid apes and humans.

see also...

Leakey

Prokaryotes

Any organism in which the genetic material (DNA) is not enclosed by membranes to form a nucleus can be called a prokaryote (*pro* = 'before'; *karyote* = 'nucleus'). Prokaryotes comprise mainly bacteria and blue-green algae and probably were the ancestors of the eukaryotes.

It has often been said that the division between the prokaryotes and the eukaryotes is the greatest single breakthrough in evolutionary history because the differences in their cellular structure are so great. The prokaryotic cell has no nuclear membrane surrounding its genetic material. Instead, a single, circular chromosome lies free in the cytoplasm. In addition, many of the functions that are performed by distinct organelles in the eukaryotic cell are carried out in the prokaryotic cell on infoldings of the membrane that surrounds the whole cell. Although they lack most of the organelles found in eukaryotic cells, prokaryotes do have ribosomes, which are assembly points for protein molecules.

Most authorities think that life first appeared on Earth about 4000 million years ago. The earliest known fossilized prokaryotes were found in Western Australia in 1980. They have been dated to about 3500 million years old and consist of at least five different types of living things, elongated and strand-like, resembling some bacteria that we know today. These fossilized remains are called stromatolites and indicate that 3500 million years ago life was already here on Earth.

It is likely that modified prokaryotes were captured by the first ancestors of eukaryotic cells and were able to live inside them in a relationship which gave mutual benefit (endosymbiosis). Organelles such as mitochondria (where most energy is released) are thought to be modified bacteria. Blue-green algae could have become chloroplasts.

One fact that supports the endosymbiotic theory is that both mitochondria and chloroplasts are organized along similar lines to prokaryotes.

see also...

Eukaryotes

Protobionts

This name is derived from *proto* (= 'first') and *biont* (= 'life form'). Protobionts are specialized droplets with internal chemical characteristics that differ distinctly from those of the external surroundings and were probably the precursors of prokaryotes. There is a huge gap between the production of individual organic molecules such as amino acids with potential biological significance and their assemblage into the integrated system which we regard as the basic unit of life – the cell. In order to bridge this gap there are two essential requirements.

1 The concentration and assembly of appropriate molecules in an orderly manner.
2 The development of a method of self-replication, the dominant feature that distinguishes living from non-living things.

Several explanations have been put forward to account for the concentration and aggregation of molecules from the dilute 'soup' that must have existed for millions of years at the Earth's beginning. The most plausible of these is based on the presence of polymers; large molecules formed by joining two or more similar molecules. These are frequently produced on the surface of water after experiments such as those of Miller, their number increasing exponentially with time, one serving as a focus for the assembly of others. These may be held together by electrostatic (positive and negative) forces and thus form coacervate droplets. Such droplets consist of an inner cluster of colloidal molecules surrounded by a shell of water molecules. As a result there is clear demarkation between the colloidal protein mass and the water in which it is suspended. Coacervate droplets may have led to the first protobionts or 'protocells'. They are, however, far removed from cells as we know them.

An important deficiency of protobionts is any mechanism for the transport of genetic information – and hence, by implication, for replication.

see also...

Origin of life; Prokaryotes

Pterosaurs

Until the Triassic period (245–202 million years ago), the only animals to become airborne were the insects. Gliding reptiles appeared when a number of lizard-like animals developed membranes supported by extended ribs. It is likely that these adaptions gave them an advantage in escaping the small carnivorous lizard-hipped dinosaurs of the same period.

The reptilian masters of the sky were the pterosaurs, which, like the dinosaurs, were descended from the archosaurs. Their ancestry can be traced through a 30 cm gliding reptile from the Triassic deposits in central Asia, *Podopteryx*. This creature had a large membrane attached to its elongated hind limbs and tail.

The best known examples of pterosaurs are the small pigeon-sized reptiles that flew by means of a pair of leathery wings which were attached to their bodies, hind legs and fore limbs. They could fly relatively efficiently, altering the wing area by stretching and folding their fore and hind limbs to manoeuvre. They probably were fish eaters, snatching them from the surface of water. To enable the pterosaurs to sustain their active lifestyle, they were warm blooded, and traces of hair or fur have been found among their fossils. The best known large pterosaur is *Pteranodon*, found in USA in late Cretaceous limestone in 1872. It had a wingspan of 6 m. In 1975 a gigantic species, *Quetzalcoatlus*, was found in Texas, with an estimated wingspan of 11–15 m. It was probably the largest flying animal ever to have soared through the skies of our planet; it would have been at least three times as big as any flying bird and more like an aircraft than any animal living before or since.

The Solnhofen Limestone of southern Bavaria in Germany has been the best source of fine pterosaur fossil skeletons, ever since the first example was found there by Cosmo Alessandro Collini in 1784. This particular type of limestone was deposited in shallow seas, just offshore of a land mass covering most of northern Europe, where mud built up over thousands of years and trapped a variety of life.

Punctuated equilibria

One of the most important developments in evolutionary theory at present concerns the idea of punctuated equilibria, first put forward by N. Eldridge and Stephen Jay Gould in 1972. These scientists pointed out that many new species begin suddenly in the fossil record, while other fossil lines remain unchanged for millions of years. However, the suddenness in geological time is very different from suddenness in genetic time. Research by A.R. Templeton in the late 1970s into speciation via genetic studies provided evidence that the splitting of a lineage can take place in tens rather than thousands of generations and by a relatively small number of gene differences.

In the fossil record that was available to Charles Darwin there were great gaps, but also there was frequently sudden appearance of many new forms at the same geological time. Evolution appeared to be happening in 'fits and starts'. Long periods of time in which certain species were present in a relatively unchanged from were interspersed by periods in which other species suddenly appeared, thereafter to remain relatively unchanged for a long time. The long periods were punctuated by short periods of change. This episodic, rather than gradual, change was named punctuated equilibrium but was in contrast to orthodox nineteenth-century Darwinism, which stressed the importance of gradualism in evolutionary change.

Speciation in the evolution of mammals was rapid, with radiation extending to around 100 families in about 30 million years. By contrast, the bivalve molluscs have evolved much more slowly, taking about 500 million years to attain the same level of diversity as the mammals. Some classic investigations were carried out by T.S. Westoll in 1949 on fossil lungfish, which have shown very little speciation because of their slow evolution. By measuring various characteristics of each fossil genus, Westoll showed that from the mid-Palaeozoic era (about 400 million years ago) change in external appearance was rapid for about the first 100 million years but that thereafter it slowed down abruptly, and for the last 150 million years, hardly any change has taken place.

Recapitulation evidence

The similarities of embryological development among vertebrates were intensively studied during the second half of the nineteenth century. In particular, the German embryologist Karl Ernst von Baer (a contemporary of Charles Darwin) was one of the first to link embryological development to evolution. He commented on a collection of unlabelled specimen jars containing preserved vertebrate embryos, observing that he was not able to tell to which class they belonged. They might have been lizards, birds or mammals, so complete was the similarity at that stage in their development. This study led to the conclusion that the embryonic development of an individual repeated the evolutionary history of the group to which the individual belonged. Thus, it was thought to be possible to trace the evolutionary history of a species by a study of its embryonic development.

About 50 years after von Baer's early observations, another German biologist, Ernst Haeckel suggested the Law of Recapitulation. His idea was summarized as 'ontogeny recapitulates phylogeny', which means that the development of the individual repeats the evolution of the phylum. It was almost like saying that, during embryonic development, an animal ascends its own evolutionary tree.

In the early stages of development the embryos of all vertebrates show remarkable similarities, hinting at a common ancestor. These similarities extend not only to external features but also to important internal structures such as the heart and arterial system. However, today the idea of embryonic resemblances is viewed with caution. While a certain amount of recapitulation is unquestioned, older ideas that a human passes through fish, amphibian and reptilian stages during early development is considered to be an over-simplification of the complex embryology of higher animals.

see also...

*Biogeographical evidence;
Comparative anatomy; Fossil
evidence; Haeckel*

Reproductive isolation

The separation of populations within a species prevents individuals breeding outside one particular population and hence prevents the mixing of genes and consequent variation. If such isolation is maintained, the species slowly diverges from its non-isolated members, as it becomes adapted to its own environment. Eventually it becomes physically different from other populations and may become incapable of breeding with them, even if contact is subsequently re-established. At this stage the isolated population has evolved into a new species.

Reproductive isolation refers to the inability of members of one population of a species to interbreed with another. It ensures that the genes from one group will not combine with those of another; thus each can accumulate its own genetic distinctions. There are several ways that groups can be reproductively isolated besides geographically.

★ **Behavioural isolation**: Courtship behaviour of two groups may keep them from breeding even where their ranges overlap. They may not recognize each other's mating signals or plumage patterns.

★ **Mating isolation**: Some species, especially insects, have very complex 'lock and key' mating organs. Copulation may be impossible because their genitalia cannot fit together. Pollen grains of flowering plants will fit only into correspondingly shaped grooves or depressions on the female parts of flowers.

★ **Hybrid isolation**: In cases where individuals of two species do interbreed, the embryo may fail to develop because the mismatched chromosomes impede the normal processes of cell division which lead to growth. Such hybrids are eliminated from the population before contributing to a gene pool. In some cases, hybrids may survive and even be fully fertile, but reduced viability means that they are unlikely to reach sexual maturity.

see also...

Allopatric speciation;
Geographical isolation

Selection

Natural selection is a process that reduces the proportion of organisms with low relative fitness within a population and which, therefore, must increase the proportion of organisms that are nearer the optimal phenotype for the specific environmental and competitive conditions. A high survival rate is of importance in that it gives the organism a higher chance of reproducing and therefore contributing its genes in the gene pool. Most phenotypic traits are controlled by many genes and tend to show a Normal distribution in the population. The position of the optimal phenotype will determine the basic type of selection. There are three basic types of selection that result in aiming at the best result for the population in a particular set of conditions. These are:

1 Stabilizing natural selection. This favours the mean at the expense of the two extremes of the distribution.
2 Directional natural selection. This favours one extreme of the phenotypic range.
3 Disruptive natural selection. This favours both extremes at the expense of the mean.

Although we can categorize selection in this way, we must remember that the classes are all aspects of the same process. The very fact that some organisms have a lower relative fitness (that is, they have lower survival and/or reproductive rates than others) means that those which keep the population going are competing with organisms that are becoming more and more similar to them. If directional natural selection gradually eliminates one extreme of the phenotypic range, then eventually there will be a random distribution around the optimum phenotype. Stabilizing natural selection will then tend to maintain this.

The first type of selection can be illustrated by heights in humans. If being of intermediate height is especially important, then those who are taller or shorter will suffer some disadvantage. This is called high selection pressure.

see also...

Artificial selection in animals;
Artificial selection in plants;
Natural selection

Speciation

There are almost as many definitions of the term 'species' as there are different species. Problems of definition arise because, in taxonomic terms, a species means organisms which share specific morphological characters, and they are often described from a few dead 'type specimens'. In nature, however, species vary in space and time, and there are often gradations in form leading to uncertainty about where one species ends and another begins. The song sparrow, *Melospiza melodia*, for example, is widely distributed throughout the USA and there are many local races which each have their own distinctive plumage and song pattern. They are still members of the same species because, just as in the case of humans. Wherever individuals from different geographical races come together, they will interbreed and give rise to intermediate populations. It is this capacity for interbreeding which unites organisms into species and which separates one species from another. Races will interbreed to produce fertile offspring but species will not. The most useful definition of a species used by biologists was formulated by the renowned zoologist Ernst Mayer:

A species is a group of actually or potentially interbreeding populations that is reproductively isolated from other such groups.

This definition obviously cannot apply to self-fertilizing organisms or to forms which reproduce solely by asexual means. Nevertheless, such groups are frequently called species where they consist of individuals that are very similar to one another. The species is the largest group of organisms that share a gene pool. There are about 3 million such groups on Earth at present, but their numbers and kinds are slowly evolving all of the time. Some of these groups are in a stable relationship with their environment, some are becoming extinct, and others are in the process of evolving new reproductive groups – that is, undergoing speciation.

With the development of new genetic technologies, such as genetic profiling, determining the make-up of a common gene pool by analysis of genes eliminates the more subjective methods of taxonomy based on external features.

see also...

Genetic profiling

Spontaneous generation

Spontaneous generation was the erroneous belief that present-day organisms can be formed from inorganic material, given the right conditions. The idea was accepted by some biologists until the end of the nineteenth century. The origins of such a belief extend back to the classical times of the ancient Greek philosophers. Aristotle (384–322 BC) is among the greatest of all biologists and no other writer has ever had greater influence on science. A profound and original thinker, he speculated on the nature of life and believed in spontaneous generation. Statements in his *Historia Animalium* showed that he mistook populations of small fish (mullet) and eels in muddy ponds as being derived from mud itself. The authority of his work was undisputed and his errors perpetuated. For several centuries, Aristotle's beliefs were accepted as fact – so, perhaps, it is not surprising that the idea of spontaneous generation was so widespread among educated people. To doubt the Aristotelian doctrine was to defy the evidence of the processes of reason and, what was much worse, to challenge the constituted religious authorities of the time. It was rare for scholars to observe and experiment for themselves until much later.

In 1652, Jean-Baptiste van Helmont published a book in which he stated that if wheat grains and a dirty shirt were put in a pot mice would be formed from the interaction of the wheat grains and the dirt in the shirt. This seems strange today but other, familiar, observations were often explained in the same way – the appearance of maggots in meat left to rot was thought to be a common example of spontaneous generation.

At least one biologist questioned these explanations: Francesco Redi (1626–97). He carried out investigations which compared the results of leaving fresh meat to rot in sealed and unsealed containers. He demonstrated that maggots were formed only if flies were allowed access to the meat. He concluded that 'if living causes are excluded, no living things arise.'

see also...

Biogenesis

Sympatric speciation

Sympatric speciation occurs within an existing population (*sym* = 'together'; *patric* = 'land'). It is less common than allopatric speciation and involves the formation of two species from one continuously interbreeding population.

Among flowering plants new species can arise by the interbreeding of existing species. This discovery surprised researchers because animal hybrids are usually sterile due to the inability of dissimilar chromosomes to carry out the normal process of sex cell formation. Hybrid plants do not have the same problem with their chromosomes because plants with very different appearances may be very similar genetically. Where such genetically similar species overlap, there may be extensive production of hybrid populations. Surprisingly, such hybridization does not seem to result in the breakdown (through merging or blending) of either parent species. This may be because hybrid populations find their own niche, interacting with the environment differently than the parent groups, thereby becoming truly a new and distinct species.

Plants can also form new species in a quite dramatic way, where whole sets of chromosomes become doubled (or even doubled again). The condition is called polyploidy (*poly* = 'many'; *ploidy* = 'set of chromosomes'). Really this is a type of mutation, where chromosomes double up by failing to separate properly during the normal process of sex cell formation. It can result in tetraploid (*tetra* = 'four') plants with four complete sets of chromosomes in each cell. Hybridization and polyploidy can create instant sympatric species of plants ready to be worked upon by selective pressures of the environment. This versatility helps to explain how flowering plants arose so suddenly and how they then so quickly spread out over the land to become the amazingly diverse group that they are today.

Sympatric speciation in animals is much rarer than in plants.

see also...

Allopatric speciation;
Allopolyploidy; Autopolyploidy;
Mutation; Niche

Variation – environmental causes

Non-genetic factors may have just as great an influence as genes on some characteristics. Heredity determines what an organism *may* become, not what it *will* become: a plant may inherit genes for tallness but may not grow tall in a poor soil. An organism's phenotype (outward appearance) depends on a combination of inherited characteristics and the effects of the environment.

Here is a simple illustration. Imagine a litter of Alsation puppies – they all have the same genes from the same parents. If we follow and compare the growth of these puppies in their separate homes, we might see that one out of the litter becomes less strong and not as large as the rest. The reason could be its genes, or it could be that it has not been given the correct diet and exercise. In other words, the effects of the environment may well have influenced its phenotype. By altering the environment we can alter the phenotype. On the other hand, a litter of Corgis all kept together will grow up to be smaller dogs than Alsations, no matter how well fed they are. Any small difference in size is then likely to be genetically determined. The genes determine the phenotype if the environment is kept the same.

The effect of diet on phenotype in humans has been well documented for many years. As long ago as the 1950s, reliable data on diet and phenotype was recorded in a study of Japanese children. Two groups of boys were raised in different environments; some were born and raised in USA where a better diet was available, the others were brought up in Japan (between 1900 and 1952, the diet and standard of living in Japan improved markedly). The American-born boys were taller at all ages than the Japan-raised ones. The improved diet and standard of living in Japan resulted in an increase in height between 1900 and 1952. The same has happened in Britain – children are now taller and heavier for their age than they were 60 years ago.

The voyage of the Beagle

At 22 years old, Darwin applied to the equally young Captain Fitzroy for the post of naturalist on board HMS *Beagle* for its epic voyage, which took place between 1831 and 1836. Fitzroy had been placed in command of the *Beagle* in 1829 after its previous commander had committed suicide. Darwin was accepted on board, having been funded by his uncle, Josiah Wedgwood (of pottery fame and wealth).

After being driven back twice to harbour by heavy seas, on 27 December 1831 the *Beagle* finally weighed anchor and set out on a journey which was to be the spark for a flame that would set the scientific world alight. The 25 m, 242-tonne solid barque-rigged brig creaked its way towards South America. It was then that Darwin's worst fears materialized. The voyage soon changed to a nauseous routine of tough shipboard life and endless seasickness. He sometimes spent whole days below deck.

At long last the *Beagle* reached South America and headed down the coast on the first leg of its voyage – sailing past the coasts of Brazil and Argentina, weaving through the terrible pounding gales of Cape Horn, and finally turning northward along the desolate coasts of Chile and Peru. Fortunately there were periods of respite, when the *Beagle* dropped anchor to put foraging parties ashore. Darwin wasted no time getting to land. He was irresistibly drawn to these new places, which harboured all manner of new and fascinating living forms. He ventured far inland, collecting species new to science, including many fossils.

However, his greatest contribution to science came as a result of his visit to the Galapagos Islands. It was here that Darwin conceived ideas of natural selection, which would eventually give rise to evolutionary concepts known today as Darwinism. On leaving the Galapagos Islands in 1835, the journey continued across the Pacific to Australasia, then across the Indian Ocean, and finally back to Britain.

> ## see also...
>
> *Darwin; Darwinism; The Galapagos Islands; Natural selection*

Wallace, Alfred Russell

Born in Usk, Gwent, in Wales, Alfred Russell Wallace (1823–1913) had little formal education. He worked in various jobs until he became a schoolmaster at the age of 21 in the Collegiate School, Leicester. Already a keen naturalist, he met H.W. Bates, a famous entomologist, at the school and in 1848 persuaded him to join in a collecting expedition to the Amazon. The expedition was successful but on the return journey their ship was destroyed and most of their collected specimens were lost at sea.

In 1854 Wallace went to the Malay archipelago, where he remained for eight years. His essay *On the Law which has Regulated the Introduction of New Species*, written in Sarawak in 1855, was a pre-Darwinist contribution to the theory of evolution but attracted little notice at the time, except from Darwin himself. After reading Malthus' *Principle of Population*, Wallace suggested the theory of natural selection, which was to become the basis of Darwinism and was being developed by Charles Darwin at around the same time. Wallace wrote an essay *On the Tendency of Varieties to Depart Indefinitely from the Original Type* in 1858 and sent it to Darwin for an informed opinion. Darwin commented: 'I never saw such a striking coincidence. If Wallace had my sketch of 1842 in front of him he could not have made a better short extract'. Darwin was prepared to concede prior publication to Wallace but, in the event, both papers were jointly read to the Linnean Society in 1858. Wallace's modesty led him to play a subordinate role to Darwin, whose theories he considered in his next book, *Darwinism*, in 1889.

Wallace's main interest, the geographical distribution of animals, was based on his observation that the distribution of animals in Malaya was divided into two distinct groups: Oriental on the west and Australasian to the east of the Bali Strait. His *Geographical Distribution of Animals* (1876) was the first synthesis of all the evidence that was available at the time and is the basis of modern zoogeography.

see also...

Darwin; Natural selection

Warfarin resistance

Warfarin is an anticoagulant poison widely used as a rodenticide for the control of rats and mice. It was first introduced in 1950 and became a popular poison against rats in particular because of its low toxicity to farm animals. It acts by interfering with the way in which vitamin K is used in the complex biochemistry of blood clotting. When the bait is administered to sensitive rats, the capillaries become much more fragile than normal, the blood fails to clot, the affected animals experience severe haemorrhages and slowly bleed to death.

Resistant strains of rats first evolved in Scotland in 1958. By 1972, resistant rats had appeared in 12 other areas of Britain. The basis of the evolution of a new strain can be explained in terms of a single gene mutation in which resistance is conferred by a dominant allele. The blood clotting process of the resistant strains is insensitive to warfarin; clotting takes place in the normal time, but there is an enhanced requirement for vitamin K. One resistant population that has been well monitored is in an area near Welshpool in central Wales, and extends to the border of Shropshire in England. Resistant rats were first noticed in Welshpool in 1959. Thereafter they were seen to spread out from the original site at a rate of three miles per year, which is the normal rate at which rats invade new territory. Their progress was dependent upon the continued use of the poison, and the constant selection pressure that was applied to the population, for several years. The environment of the rats, in terms of diet, was suddenly and deliberately changed by the intervention of human activity. This altered the selective value of the genes normally involved in blood clotting, and their action suddenly became lethal in the homozygous recessive condition. A few mutants carrying the dominant resistance allele were then favoured by selection. In the new environment, the genetic composition of the population changed. This is an example of selection exerting its effect remarkably quickly.

see also...

Selection

Further reading

Brookfield, A.P., *Modern Aspects of Evolution* (Hutchinson, 1986).

Broom, Robert, *The Coming of Man: Was It Accident or Design?* (Witherby, 1933).

Dart, Raymond A, with Dennis, Craig, *Adventures with the Missing Link* (Sustitutes Press, 1959).

Darwin, Charles, *The Origin of Species* (John Murray, 1859).

Darwin, Charles, *The Descent of Man* (John Murray, 1871).

Darwin, Francis, *The Life of Charles Darwin* (John Murray, 1902).

Dawkins, R., *The Selfish Gene* (Oxford University Press, 1976).

Dawkins, R. *The Blind Watchmaker* (Longman Scientific and Technical, 1986).

Dowdswell, W.H., *Evolution: a modern synthesis* (Heinemann Educational, 1984).

Fortey, Richard, *Life An Unauthorised Biography* (HarperCollins, 1997).

Gould, Stephen Jay, *Wonderful Life: The Burgess Shale and the Nature of History* (Hutchinson Radius, 1989).

Jenkins, Morton, *Teach Yourself Genetics* (Hodder & Stoughton, 1998).

Jenkins, Morton, *Teach Yourself Evolution* (Hodder & Stoughton, 1999).

Leakey, Richard, *The Making of Mankind* (E.P. Dutton, 1981).

Leakey, Richard, *The Origin of Humankind* (Weidenfeld & Nicolson, 1994).

Leakey, Richard and Lewin, Roger, *The Sixth Extinction: Biodiversity and its survival* (Weidenfeld & Nicolson, 1996).

Lewin, Roger, *The Origin of Modern Humans* (W.H. Freeman, 1993).

Miller, S.L. and Orgel, L.E., *The Origin of Life on Earth* (Prentice-Hall, 1974).

Oparin, Alexander, *The Origin of Life on the Earth* (Macmillan, 1938).

Rose, Steven, *The Chemistry of Life* (Penguin, 1991).

Rose, Steven, *Lifelines* (Penguin, 1997).

Smith, John Maynard, *The Theory of Evolution* (Penguin, 1975).

Smith, John Maynard, *Did Darwin Get it Right?* (Penguin, 1993).

Walker, Alan and Shipman, Pat, *The Wisdom of Bones: In Search of Human Origins* (Weidenfeld and Nicolson, 1996).

Wallace, Alfred Russell, *Darwinism* (Macmillan, 1898).

Also available in the series